The Story of Donald Simpson Bell VC

The Story of Donald Simpson Bell VC

Football's War Hero

Iain McMullen

Pen & Sword
MILITARY

First published in Great Britain in 2026 by
Pen & Sword Military
An imprint of Pen & Sword Books Limited
Yorkshire – Philadelphia

ISBN 978 1 03611 902 7

A CIP catalogue record for this book is available from the British Library.

Typeset by Mac Style
Printed in the UK by CPI Group (UK) Ltd, Croydon, CR0 4YY.

The Publisher's authorised representative in the EU for product safety is Authorised Rep Compliance Ltd., Ground Floor, 71 Lower Baggot Street, Dublin D02 P593, Ireland.
www.arccompliance.com

For a complete list of Pen & Sword titles please contact:

PEN & SWORD BOOKS LIMITED
47 Church Street, Barnsley, South Yorkshire, S70 2AS, England
E-mail: enquiries@pen-and-sword.co.uk
Website: www.pen-and-sword.co.uk
or
PEN AND SWORD BOOKS
1950 Lawrence Road, Havertown, PA 19083, USA
E-mail: uspen-and-sword@casematepublishers.com
Website: www.penandswordbooks.com

To those who fell and those left behind.

Contents

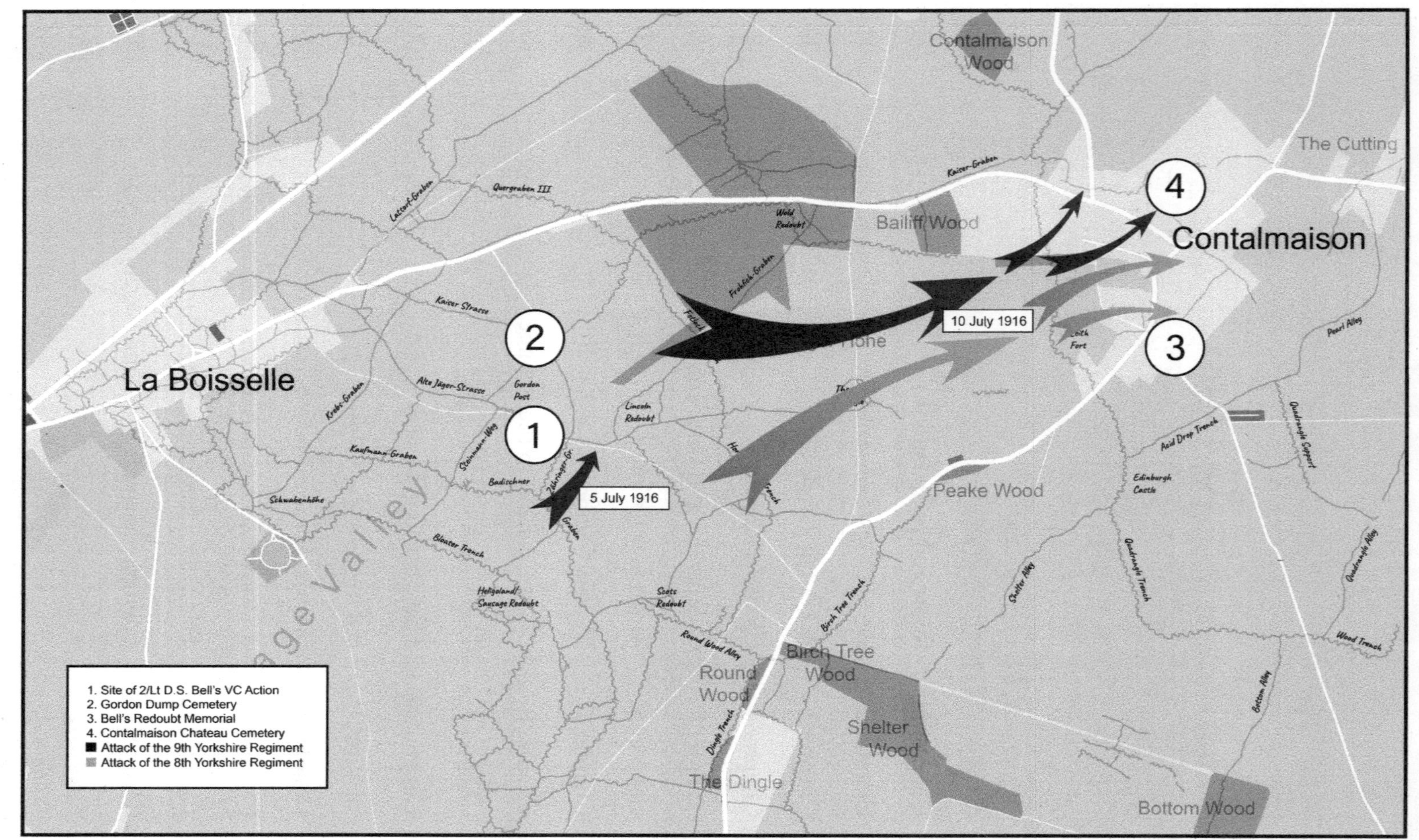
La Boisselle
Contalmaison
Contalmaison Wood
The Cutting
Bailiff Wood
Peake Wood
Birch Tree Wood
Round Wood
Shelter Wood
The Dingle
Bottom Wood
10 July 1916
5 July 1916
Kaiser-Graben
Quergraben III
Lattorf-Graben
Kaiser Strasse
Alte Jäger-Strasse
Gordon Post
Krebs-Graben
Kaufmann-Graben
Steinmann-Weg
Badischner
Schwabenhöhe
Bloater Trench
Heligoland/ Sausage Redoubt
Scots Redoubt
Round Wood Alley
Dingle Trench
Birch Tree Trench
Lincoln Redoubt
Wold Redoubt
Fricourt-Graben
Leith Fort
Acid Drop Trench
Edinburgh Castle
Quadrangle Support
Quadrangle Trench
Quadrangle Alley
Shelter Alley
Pearl Alley
Wood Trench
Bottom Alley
1. Site of 2/Lt D.S. Bell's VC Action
2. Gordon Dump Cemetery
3. Bell's Redoubt Memorial
4. Contalmaison Chateau Cemetery
Attack of the 9th Yorkshire Regiment
Attack of the 8th Yorkshire Regiment

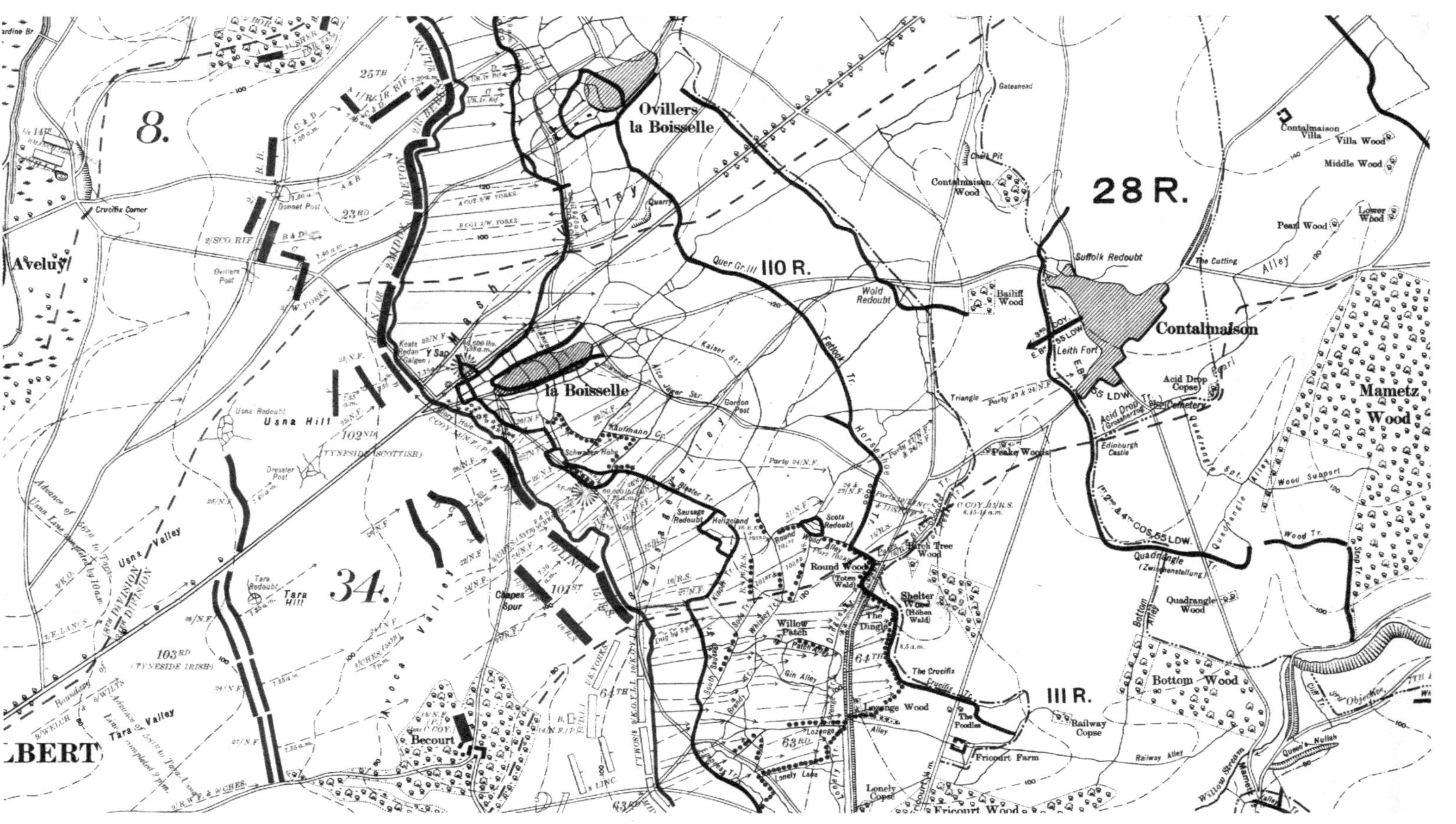

Strategic map of the battlefield on 1 July 1916. (*Author's Collection*)

Foreword

Only through investigating personal experiences in depth can we truly get to understand what it was like for those unfortunate enough to be caught up in the Great War. Iain McMullen has done us all a great service by exploring the life 'in the round' of Donald Simpson Bell, a man who achieved some renown as a footballer, but also became an acclaimed hero considered worthy of being awarded a Victoria Cross. Instead of a 'cartoon' image of a hero, defined solely by a few moments of monumental courage, we get a fully developed picture of the man and his background in Harrogate. When I was employed interviewing veterans for the Imperial War Museum back in the 1980s, I always tried to look at a man's family, his social circumstances, his hobbies, his education, what he was good at and sometimes - just as important - what he was not so good at! In that way you got a picture of the man *before* the war and could better judge their reactions to the stresses and challenges of that epoch defining event.

Bell trained as a teacher at the Westminster Training College where his letters show him to have been an amusing correspondent. Here he cemented his reputation as a multi-talented sportsman before he began teaching in Harrogate in 1911. His parallel career as a footballer was also beginning to take off as he was scouted as a full back, signing an amateur contract and occasionally playing for the reserve team of the mighty Newcastle United. His opportunities were limited so as an amateur he also played for several local teams including Bishop Auckland. Then another step forward as he signed a professional contract with Bradford Park Avenue, where he was on the fringes of the first team when they were promoted to the First Division in 1914. The match reports quoted are fascinating, covering the action on the field rather than the personality driven psychodramas and VAR controversies of today. One very real controversy was the decision of the Football League not to suspend the 1914/15 season following the outbreak of war. The

Bradford Park Avenue players and club threw themselves into fundraising for the war effort, and even underwent some voluntary military training to prepare them should they be required to enlist. It was not enough and Bell requested that his contract be terminated and on 28 October 1914 he duly enlisted as a private in the West Yorkshire Regiment. Behind him the season stumbled on, but it would be the last until the war was over.

Bell underwent the usual basic training, but was persuaded to apply for a commission which he received in June 1915. He attended an officer training course, but there was surely never any doubt that he was ideal officer materiel – intelligent, brave, strong and determined. After a frustratingly long training period with the 11th Yorkshire Regiment, he was at last despatched to France in November 1915. Following a brief period at the Étaples 'Bull Ring', Bell was posted to the 9th Yorkshire Regiment, part of the 69th Brigade, 23rd Division. His training was not yet complete – there was much more to learn to hammer home the practical skills of trench warfare which would soon be life or death. He was also carefully introduced to the front with a series of working parties before they he into the line in December 1915.

Now we discover that detail is not boring! What is boring are bland, unfocussed accounts spouting cliches that don't get to grips with what was really happening! McMullen excels in providing us with the nitty-gritty of the military situations that Bell and the 9th Yorkshires encountered with welcome commentary from Bell's lively letters. It is pleasing to record he also managed to continue to play football, even securing a new ball for the men from his old Bradford Park Avenue club. He also got married to Rhoda Bonson whilst on leave on 5 June 1916. They would not spend much time together.

All roads lead to the Somme and Bell, with newly sheared close cropped hair, was all-unknowing moving towards the denouement of his life. They moved into the front line on the night of 3-4 July. Next day they were ordered to attack the hotly contested Horseshoe Trench. Seizing the initiative, Bell lead a heroic attack on an enfilading German machine gun which he put out of action, before clearing the nearby dugouts and the objective was secured. His own account was modest, but the deed merited the award of his VC. But Donald Bell would never know that – just a few days later on 10 July 1916 he was killed in an attack on the German fortress of Contalmaison.

Demonstrating incredible courage, he led attacks to help consolidate their grip on the village, fighting on despite being wounded several times, until the fatal blow struck him down. He was a hero - and a likeable one at that. That is not the end of the story as his wife, his family, and the wider Harrogate and football community had to cope with their loss, one of tens of thousands, but far closer to home.

Peter Hart, 2026

Preface and Acknowledgements

At its most basic level, this biographical study recounts the life and death of Donald Simpson Bell VC, the only English professional footballer – and one of only two British professional players – to be awarded the Victoria Cross during the First World War. It is a story familiar to many, yet one that has not escaped the distorting effects of time. For more than a century, it has been shaped as much by romanticized half-truths as by fact and further muddied by historical inaccuracies, leaving a remarkable story only partially told.

At the heart of this book lies a determination to examine Bell's life in its entirety, set within its wider historical context, as he lived it. Not to diminish his courage, but to restore depth and dimension to a legacy too often reduced to two isolated acts of battlefield gallantry in July 1916. While his VC action on 5 July, and his death five days later, remain defining moments, they represent only part of a far richer and more nuanced portrait of a man driven by principle and integrity, who was also a devoted son, brother, husband and friend.

I am grateful to Peter Hart for agreeing to write the foreword to this book. His knowledge of the First World War and his longstanding work in the field add valuable context to the story that follows.

I have endeavoured to place Bell's story within a broader historical framework, examining the relationships and influences that shaped his life. From his formative years in Harrogate, where faith and family served as the centripetal force, through his two years at Westminster Training College, his return to Yorkshire as a schoolteacher, finding modest fame on the football field, and finally, his transition to life in uniform and active service in France.

Fortunately, Don was a prolific letter-writer, and the hundreds of pages of correspondence preserved in the Bell family archives provide a vivid and often intimate insight into his life, his hopes and his aspirations.

This is not a book about football, nor is it a conventional military history, though both subjects are covered in depth. Don's war did not unfold in isolation; it was an experience shared with the officers and men of his battalion. Therefore, I have sought to chart his service within the wider collective narrative, particularly during the fighting at Horseshoe Trench and Contalmaison. By affording these operations the level of detail they demand, I hope to have produced a distinct and substantive record of the 9th (Service) Battalion, Alexandra, Princess of Wales's Own (Yorkshire Regiment) on the Somme – a history that has, hitherto, been largely neglected.

I have also explored the significance and the broader ramifications of the contentious 1914/15 English football season, when the decision to continue official competition, despite mounting criticism, saw the game and those who played it being attacked and vilified in a bitter, highly public war of words played out across the pages of the national press.

It is important to note that the story does not end with Don's death or the award of the Victoria Cross. His life was one rooted in family, and it would be remiss of me not to consider the impact of his loss on those closest to him. By tracing the course of the war to its conclusion, and into the decades that followed, I have sought to bring the story full circle, offering some measure of closure to a life cut tragically short, like so many others, in its prime.

Of course, this task would have been significantly more difficult without the assistance of others, and this book owes much to many people. From the outset, I have received a level of encouragement and support far surpassing all reasonable expectations.

First and foremost, I owe a profound debt of gratitude to Don's family. Their trust, generosity and willingness to help have been unwavering throughout. In particular, I offer special thanks to the grandchildren of Dolly Angus (née Bell). My deepest gratitude must go to Janet Bates, who not only welcomed me into her home but also granted me the privilege of accessing the family archives – a truly humbling experience. Also special thanks to Kate Bowman, who answered countless questions about her grandmother and great-uncle Don with patience and genuine interest, as well as Malcolm and Ian Dawson, whose insights and recollections of Dolly and their family were equally invaluable.

My thanks also go to Richard Bell, grandson of Billy Bell, who generously shared his personal archive with me. His receptive response to my first, tentative email marked the beginning of my connection with the family. Finally, I sincerely thank Alison Booth and Ruth Umpleby for sharing the stories of their grandmother, Gertie, and for their invaluable assistance with photographs and details about the extended family.

Elsewhere, Steve Erskine and Zoë Utley at the Green Howards Museum in Richmond deserve special mention, as do Peter Forsaith and Tom Dobson at the Oxford Centre for Methodist and Church History. Their generosity, help and support during research visits was both steadfast and invaluable.

I must also express my gratitude to the late Richard Leake, whose early research on Bell and fellow VC recipient Archie White served as a foundation and opened the door to their stories.

Thanks go to the staff at the British Newspaper Archive, the Commonwealth War Graves Commission, the Green Howards Museum, Harrogate Grammar School, Harrogate Library, the Imperial War Museum, Manchester Library, the National Archives, the National Football Museum, the North Yorkshire County Records Office, the Oxford Centre for Methodism and Church History, the Professional Footballers Association, Tyne and Wear Archives and York Army Museum, among many others.

In addition, I would like to thank those who have offered advice and input during the research and writing process, including Sharon Bushby, Tim Clapham, Sarah Davies, Eleanor Davison, Charles Fair, Alex Jackson, Paul Joannou, Nicholas Melia, Jonathan Porter, Abby Pullen, Daniel Reed, Paul Reed, Rocky Salmon, Gary Sheffield, Dan Sudron, Julie and David Thomson and Judith Yeats.

At Pen & Sword, I am especially grateful to Harriet Fielding, who has answered a multitude of questions and queries along the way and shown incredible patience and diligence throughout the editorial process. My thanks also go to Lisa Hooson, who set things in motion following my initial proposal.

I am also indebted to the Western Front Association, the Long, Long Trail, the English National Football Archive and the many contributors to the Great War Forum, whose shared research and extensive knowledge have proven invaluable.

My preoccupation with the First World War began almost four decades ago, and I owe a great deal to my parents, who took me on my first trip to the Somme at the age of 10, thus igniting a lifelong obsession.

Finally, but most importantly, I want to thank my long-suffering wife, Natalie, and our four children, George, Grace, Amy and Lucy. Over the past twenty years, they have endured countless visits to battlefields in all weather conditions, with patience and good humour. Without their unwavering support, encouragement and understanding, this book would not have been possible, and for that I remain eternally grateful.

IM

Prologue: A Letter from France

14 December 1915
British Front Line
Bois-Grenier Sector, France

2/Lt. D.S. Bell
11th Yorks attd 9th
69th Brigade
23rd Division
B.E.F.

Dear Mr. Breare,
At the present time I am in the first line trenches, the battalion having relieved the West Yorks on Friday last. When we took possession of the trenches, we thought we were fortunate to have taken over such a dry portion of the line, but next morning we had an eye-opener. The Germans hold higher ground than we do, and several streams run from their one to ours. About eight o'clock on the Saturday morning these streams rose rapidly, and in a few moments our trenches were flooded, knee deep, and in some places taking one up to the waist. Until we recognised the danger spots, there were frequent mishaps in the way of duckings, one officer breaking the record with three in one day. Since Saturday, it has been a case of pump, pump, pump! That is, after we had dammed the streams (in more ways than one), and today the trenches are returning to their former state. The sudden influx was caused by the bursting of a dam on the German side by means of our artillery. I am afraid the latter did us a bad turn, but we are making up for it this afternoon by shelling the trenches opposite. Sandbags are flying in all directions, and 'Fritz' is having a surfeit of iron rations. This is my first spell in the trenches, but except for the difficulty of keeping dry, it has not been an unpleasant experience. I am thankful to know that we shall be out

of trenches for Xmas and consequently should be able to celebrate the day in a fitting manner.

Thanking you once again for your kindness and wishing you the compliments of the season.

Don[1]

Chapter 1

Roots and Foundations

Summerbridge is a scenic and serene village, tucked away among the rolling green dales and windswept moorland of Yorkshire's historic West Riding. It owes much of its bucolic charm to the River Nidd, which winds languidly through the valley on its southward course to Fringill Dike before sweeping east in search of the Ouse.[1] Along its banks, wildflowers grow in untidy profusion, while stands of ancient oak and birch cast long shadows across waters that teem with brown trout, roach and grayling.

Today, the river presents a scene of quiet beauty and tranquillity. But in the early nineteenth century, it was the working artery that powered a burgeoning textile trade. Mills lined its course from Pateley Bridge in the north to Summerbridge in the south, providing work for a sizeable proportion of the population. The largest and most prominent of these, New York Mill, employed more than 150 workers at its peak and stood at the heart of the local economy.[2]

This industrial growth transformed Summerbridge from a scattered cluster of stone dwellings into a thriving community. By the 1850s, the village was home to several hundred residents and could boast its own Wesley Chapel, school and public house – the Flying Dutchman. A Mechanics' Institute, with over 100 members and a 'library containing more than 1,900 volumes', was also opened in 1841. 'This educational and literary institution,' noted one contemporary account, 'speaks well for the inhabitants, who are principally employed in the adjoining manufactories.'[3]

It was during this period of rural prosperity that Donald Simpson Bell's father entered the world. Born in Summerbridge on 18 June 1859, the forty-fourth anniversary of the Battle of Waterloo, Smith Bell was one of five children and the eldest of two boys.[4] The scrawled entries on his birth certificate and the 1861 census reveal that his own father, Richard Bell, earned a living as a journeyman bobbin turner, crafting lathe-turned wooden bobbins for the mills that dominated local trade.[5]

Tragedy struck the Bell household on Christmas Eve 1866, when their youngest child, Annie – barely twelve months old – died from pneumonia following a bout of bronchitis.[6] The family were devastated by their loss, but death would return with grim regularity in the years that lay ahead. When Smith Bell was 15, his mother, Mary Ann, contracted meningococcal meningitis and fell into a coma. The 41-year-old, who had recently given birth to her second son, clung to life for almost three weeks but eventually succumbed to infection and died at home on 9 September 1874.[7] Scarcely six weeks later, the infant, George William, perished too.[8]

By 1881, Richard Bell had inherited eleven acres of land in Summerbridge, where he lived at White House Farm with his son, now 21, and two daughters: 20-year-old Sarah Jane and 11-year-old Minnie.[9] His son-in-law, Robert Petty, who had married Sarah in 1878, also lived at the property, along with their two young children and an elderly aunt.[10] That year's census found Smith employed as a butcher, a trade he would pursue for the remainder of his working life. He had also fallen in love.

On 19 June, 1883, the day after his twenty-fourth birthday, Smith Bell married 22-year-old dressmaker Annie Simpson in a ceremony at Harrogate's Wesley Chapel.[11] Annie was a local girl with strong family ties to the town. Her father, James Simpson, was an eminent local building contractor who had overseen construction of the new chapel in the 1860s, replacing the original built forty years earlier by her grandfather.[12] He would also build the impressive Church of St Michael and All Angels in Beckwithshaw, consecrated in 1886 and now a Grade II listed building. Annie's eldest brother, David Simpson, was another successful Harrogate builder and contractor. His most prominent work includes the sprawling Duchy Estate, a leafy residential quarter for the wealthy of the town; the 'splendidly Gothic'[13] Royal Arcade on Parliament Street; and the lavish Grand Hotel on Cornwall Road. Both men also held prominent positions in local government and exerted considerable influence over Harrogate's civic and business affairs for many years, each serving terms as mayor and deputy mayor. David Simpson's involvement with the town council eventually spanned more than three decades, culminating in his election as the first honorary Freeman of the Borough of Harrogate in 1923, in recognition of his 'eminent public service'.[14]

In 1885, Annie Bell gave birth to her first child at the couple's home at 8 Granville Terrace – a daughter called Ellen, known affectionately as Nellie. Minnie was born two years later, followed by William – predictably shortened to Billy – at the end of 1888. The family had now relocated across town to 1 Somerfield House on Parliament Street and stood comfortably on the middle rung of Harrogatonian life.

Annie's world was torn apart in 1890 when her sister, Ellen Simpson, died in tragic circumstances following an accident at Starbeck Station. The 25-year-old was returning from a day trip to Scarborough with her elder sister, Emma, and cousin, Aaron Waines, when, much to the horror of the other passengers on board the crowded train, the carriage door suddenly flew open, hurling poor Ellen headfirst onto the platform. She survived long enough to be carried home, but never regained consciousness and died in the early hours of the next morning from a fractured skull.[15] An inquest, held at the Exchange in Harrogate two days later, heard testimony from multiple witnesses, including Ellen's cousin, who described the harrowing moments leading up to the accident:

> 'On arrival at the latter place [Starbeck] the deceased got up to take down the umbrellas from the rack. As the train passed the level-crossing gate the deceased said, "I'll look and see if the trap is here for us." She then got hold of the strap of the carriage door, and was bidding two ladies "good-bye", when the door flew open, and she was thrown upon the platform. The train jerked; and he thought this was owing to the train stopping. It seemed to stop and then drew on again. From where the train stopped to where the deceased fell it would be a distance of twelve yards. Deceased was picked up, and said, "Oh, my head!" She never spoke afterwards. Dr. Ward deposed to being called to attend the deceased when he found her bleeding from a wound in the scalp. She was quite unconscious, and presented all the features of a fracture of the skull. On her removal home, from the general condition of the patient, the case was considered a hopeless one. She died shortly before one o'clock.'[16]

The North Eastern Railway Company were eventually found guilty of negligence and ordered to pay Ellen's grieving parents £100 in damages, but it was small consolation for the loss of their daughter.[17] Tragically, within four years, Emma Simpson would be dead too, losing a prolonged battle with diabetes at the home she shared with her husband, Albert Robinson, in Grange-over-Sands, Cumbria. She was just 30 years old.[18]

At the time of Ellen's death, Annie Bell was carrying her fourth child, and twelve weeks later she gave birth to a healthy baby boy at their home on Parliament Street. Donald Simpson Bell, born on 3 December 1890 with a shock of dark hair and bright eyes, was the couple's second son and the only one of their eight children to carry his mother's maiden name. His birth coincided with a period of sustained growth among Harrogate's middle classes. Evidence of the Bell family's rising social standing can be seen in the 1891 census, which reveals a domestic servant, 16-year-old Emily Johnson from the nearby village of Farnham, living with them.[19]

As the family's prosperity grew, so too did their brood. Gertrude, known as Gertie, was born at the start of 1892, followed by Annie – named after her mother, but typically called Nancy or Nance – in May 1894. Three years later, they welcomed another daughter, Dorothy, known as Dolly throughout her life. Four of the seven children – Minnie, Billy, Don and Gertie – were now pupils at St Peter's Church of England School, less than 200 metres from their home, while Nellie, their eldest, had recently left to find work.

Like their parents before them, Smith and Annie Bell maintained a lifelong fealty to Wesleyan Methodism, with family ties to the movement that predated Harrogate's first Wesley Chapel, opened in 1824. Religion suffused the lives of their children too, with the town's new chapel at the centre of their faith. Dominating its surroundings on the corner of Chapel Street and Cheltenham Crescent, its striking hexastyle Corinthian facade made it one of Harrogate's most recognizable landmarks. It was imposing both in size and influence and played a formative role in the lives of the Bell children, as it did for many within the town's thriving Methodist community.[20]

The family were respected members of the congregation and dedicated a considerable amount of time to the chapel's religious and musical activities. Smith Bell was appointed choirmaster at the end of the nineteenth century, a position he would hold until the 1920s, and his two sons and four daughters sang in the choir. Referred to fondly as the 'Merry Peal of Bells', they were well known in the local area, frequently travelling to villages further afield to perform concerts.[21] Each child possessed notable vocal ability, but it was Nancy, an accomplished and highly regarded contralto, who usually drew most plaudits.

Among the Wesleyans' most prominent endeavours was the Sunday school on Cheltenham Parade, where the Bell children received spiritual and moral instruction, later volunteering as teachers. Opened in 1873 at a reported cost of £3,000,[22] the school was particularly popular among working-class families, who traditionally lacked access to formal education, and it played a crucial role in shaping the religious beliefs and practices of many young people in the town.[23]

By the early 1900s, Harrogate's population had risen to more than 26,000.[24] This rapid growth drove significant development in the town, and rows of newly constructed townhouses were consuming the last tracts of open land between High and Low Harrogate. Exuding an air of genteel wealth, this evolving urban landscape presented new opportunities for the upwardly mobile Bell family. By 1901, they had left Parliament Street behind and taken possession of a commodious three-storey semi-detached house at 87 East Parade – Milton Lodge.[25]

The following year, the Bell brothers secured scholarships to Knaresborough Grammar School. Details of the awards, published in the *Knaresborough Post*, provide early evidence of Don's exemplary academic ability:

> 'Two sons of Mr. Smith Bell, of East Parade, Harrogate, have recently been fortunate in securing two West Riding County Council Scholarships. Don Bell, who is only 11 years of age, secured a scholarship in class "a", and William S. Bell (13) in class "c". These are for two years' free scholarship, and the boys are attending the Knaresborough Grammar School.'[26]

Given that scholarships were highly competitive and awarded to only a select number of students each year, this marked a significant step in the lives of both boys. They would flourish at Knaresborough, but their time there would be brief. Within three years, Billy Bell had left to seek employment, while his brother had enrolled at another school – one that would prove far more influential in shaping the trajectory of his life.

The abolition of state school fees in 1891 and the educational focus of Queen Victoria's 1897 Diamond Jubilee had shifted Harrogate's priorities toward improving public education. As part of its Jubilee celebrations, the town announced ambitious plans to build a new School of Art and Technology on land at Haywra Crescent. The foundation stone was laid in December 1898, and less than two years later, the school opened its doors to its first cohort of students.[27] Harrogate's commitment to public education was further reinforced by the 1902 Balfour-Morant Act, which placed secondary education under local authority control. As a result, Harrogate Borough Council assumed responsibility for the School of Art and Technology, where it would establish 'a secondary day school for boys and girls, under the inspection of the Board of Education and the West Riding County Council'.[28] Further details appeared in the *Harrogate Herald*:

> 'The object of the school is to prepare youths for industrial, commercial and professional pursuits. It will supply a sound general modern education, from the sixth standard of an elementary school, to the work required for matriculation at the universities. Classes from 9.00am to 12.15, and 2.00 to 4.30pm, except Saturdays. A certain amount of homework must be done by each pupil.'[29]

The British Government did not expect the secondary education clauses of the Balfour-Morant Act to be implemented until 1 April 1904, but the town was ahead of the curve. On 16 September 1903, an advertisement appeared in the *Harrogate Herald* announcing the opening of the new school:

> 'Harrogate Technical School. Session 1903–4. Commencing 21st September. The Committee have pleasure in announcing that the staff of teachers has been considerably augmented and that additional facilities are now offered to all students ... The Secondary Day School's new department will provide a thorough scientific and general education for both boys and girls from the work of the sixth standard in an elementary school to that required for matriculation in the universities.'[30]

Even so, admission to the new Municipal Secondary Day School was not a formality. Scholarship applicants were required to sit written examinations in English and Arithmetic, along with several oral tests. Surviving records show that most passed, but a small number fell short and were turned away.[31] The initial intake of students when the school first opened its doors on 21 September 1903 numbered fewer than fifty boys and girls. They were taught in four dedicated classrooms, all located on the ground floor. However, the science laboratory, art room and kitchen had to be shared with the neighbouring Art School. Physical training, a key part of the curriculum, took place in the dark and dusty confines of the basement, which also housed the Art School's clay and plaster work area – an arrangement far from ideal. It was a modest start, but these early limitations were soon overcome as the school expanded in both size and reputation in the years that followed.

Donald Bell arrived at the fledgling school in midsummer 1905. Sadly, he did so under the shadow of more family bereavement. On 5 May, her own forty-fourth birthday, Annie Bell bore her eighth child and sixth daughter, Mary. Eighteen days later, the infant went into 'convulsions'[32] and died. Little else is known, but with Alexander Fleming's discovery of penicillin still more than two decades away, Mary's death is as a stark reminder that infant mortality remained a grim but all too common companion for many families.

Fourteen years old, Don had now resolved to pursue a career in teaching, a pathway that granted him free admission to the new school until the age of 16. With satisfactory progress reports, he could then sit the pupil-teacher examination set by the Board of Education and, if successful, begin a two-year apprenticeship. To encourage further advancement, a bonus was awarded to each pupil-teacher who gained entry to a training college. The system was not without its flaws, and a series of reforms saw it gradually replaced by more formal teacher training institutions. Nevertheless, the pupil-teachership served as a crucial stepping stone in Bell's education, granting him both access to the new school and a defined route into the teaching profession.[33]

Unfortunately, documented evidence of his time at the school is sparse. However, through recollections preserved by family, friends and former classmates, we get snapshots of this formative period, including the beginning of a friendship that would profoundly influence the course of his later life.

Don first met Archie White shortly after arriving at Haywra Crescent in 1905. Though one year younger,[34] White and Bell shared much in common, and the two boys quickly formed a close bond. Both possessed a sharp, quick-witted sense of humour and excelled academically. However, it was their sporting ability that truly set them apart from their peers. They were the school's outstanding athletes, distinguishing themselves at cricket, football, rugby and athletics, and regularly appeared together for Harrogate YMCA in the local amateur leagues. Yet White himself would later concede that it was Bell who stood above all others on the sports field, establishing a reputation that would follow him throughout his life. 'Every schoolboy has a hero and Don, who was about a year older than me, was mine,' White wrote just before his death in 1971. 'He had every quality that I envied – good looks, unusual strength, an even and kindly temper.'[35]

The writer A.A. Thomson, best known for his whimsical books on cricket, also attended the school. Although several years younger, Thomson left a fascinating homage to the pair in his autobiographical novel *The Exquisite Burden*, published in 1936. Through the eyes of his protagonist, Philip Gomersal, Thomson recounts a series of fleeting yet impressionable encounters with the school's two sporting heroes, Tom Leyland and Roy Moore – thinly veiled portrayals of Bell and White. The details of this retrospective fictional account are not always reliable, nor is its chronology entirely accurate. Nevertheless, Thomson does provide a compelling insight into the respect both boys obviously commanded among their contemporaries:

> 'In common with the whole school, Philip had always frankly worshipped the great Tom Leyland [Bell] who, at least on the football field, was a schoolboy-phenomenon. Had he not played fullback for Nidvale United and was he not, at sixteen, bigger and stronger than any master? Mr. Freeth called him "Son of Anak" or "Lifeguardsman Shaw", and, indeed, he towered above his schoolmates like a friendly St. Bernard among a litter of terrier puppies. He was quiet-mannered and unassuming and looked on school-life with a grin of wide good nature. He received as much flattery and hero-worship as would have turned the heads of any ten ordinary boys, but he had in his nature a core of sane simplicity which steadfastly refused to be spoiled. He was so strong that he could afford to be gentle and his good temper was unbreakable. His one eccentricity was the wearing of blue serge suits, which in time acquired the shininess of a mirror, so that rumour said you could see yourself in his trousers'-seat. While no-one had ever made actual proof of this, it was universally felt that you had only to ask and Tom would good-naturedly bend down and submit to the experiment. Philip dumbly adored him all his days at school, and when Tom left (at the end of Philip's third year) he repaid this devotion by an act of grace, recommending him as First Eleven goalkeeper for the coming season.'[36]

Don was every inch the sportsman. Tall, strong and powerful, he cut a dashing figure even as an adolescent. He possessed a blistering turn of speed and could run the 100-yard dash in under eleven seconds, an asset that served him equally well on the rugby field. He was also notably versatile, excelling across a range of sports. He captained the school's cricket First XI and was, by all accounts, a true all-rounder: formidable with the bat and guileful with the ball. Indeed, many believed he had the attributes to reach the highest level, had he pursued that path. But it was on the football pitch that Don felt most at home, and it was there that he truly forged his reputation.

Another passage from *The Exquisite Burden* sees 'Leyland' take charge of a group of unsupervised third-formers idly kicking a football while waiting for their games teacher:

> 'Suddenly the boys around him stopped kicking and stood still in attitudes of awe. A mighty figure was striding across the ground towards the First-Eleven pitch. He wore his top-coat over his jersey and shorts, but as his coat-tails flapped to his stride, his great legs displayed themselves like the boles of forest-trees. It was not strange that small boys held their breath, for this was the great Tom Leyland himself, a genial giant and already, at sixteen, a figure of heroic legend. He bestrode the petty world of school football like a colossus, only turning out for the First Eleven on Wednesday: on Saturdays he played full-back for the town team, Nidvale United. His physical strength was enormous, so was his cheerful good nature. He was as dazzling in his football prowess as any hero in Philip's Sunday School prizes, for he was accounted an outstanding full-back even in the West Riding Senior League.'[37]

Beyond school, the Wesley Chapel remained central to Don's life, and he was now dedicating some of his free time to teaching at its Sunday school. He was also finding modest fame in local footballing circles, turning out regularly for Starbeck AFC in the West Yorkshire League, much to the chagrin of his father. Smith Bell regarded his son's athletic pursuits as a frivolous and unnecessary distraction from his academic responsibilities, and they remained a source of friction between the pair for many years.

In October 1907, Don's eldest sister, Nellie, married 27-year-old Henry Jackson of Leeds at the Wesley Chapel. Henry's father, William Jackson, was the proprietor of the Jubilee Laundry on Anchor Road, High Harrogate – a business Henry would eventually inherit. The couple's first child, William, named after the eldest of Nellie's two brothers, was born in July 1908. A second son, Donald Bell Jackson, named after her younger brother, followed three years later.

Shortly after the birth of his nephew, Don matriculated at the University of London, passing examinations in English, Mathematics, Latin, French and Modern History. His results were outstanding and placed him in the First Division, one of just 185 students to achieve such distinction from more than 1,500 who sat the exams.[38] This exceptional academic performance reflected both his intellectual aptitude and personal discipline, qualities that would stand him in good stead throughout his life.

With his academic future secured, Don approached his final year of study in Harrogate with renewed vigour and enthusiasm. On 1 August 1908, he began a student-teacher placement at Starbeck School, where he quickly 'gave promise of becoming an extremely successful teacher'.[39] So successful was this year-long placement, in fact, that it laid the foundation for his eventual return to the school as a newly qualified teacher in 1911.

Chapter 2

Floreat Westminsteriensis!

> 'Floreat Westminsteriensis! Pass the word along! Here's a hand, and there's the other;
> Friendship's pledge to one another, Shake for Auld Lang Syne, my brother,
> Shout the good old song – Floreat Westminsteriensis! W's, Hurrah!
>
> – Westminster Methodist Training College Song

By late summer 1909, the grand halls and vaulted corridors of Westminster Training College had echoed with the footsteps of fledgling Methodist schoolteachers for more than half a century. Founded in 1851, the college was the principal institution for the 'training of masters and mistresses for Wesleyan Day Schools',[1] and a daunting prospect for new arrivals. Occupying a striking collection of neo-Gothic buildings on Horseferry Road in central London, close to the spot from where James II fled to France two years before his day of reckoning on the banks of the Boyne in Ireland, its scale and grandeur immediately struck students as they passed through the austere main gatehouse into the front quadrangle for the first time. The college was imposing in both scale and presence. Yet for most, it would become a place of fellowship and formation.

By the time Donald Bell arrived at Westminster in mid-September 1909, the college had been a bastion of male teacher training for over thirty years.[2] But that had not always been the case. During the first two decades of its existence, it operated as a coeducational institution, preparing hundreds of young men and women of the Wesleyan faith for service in its burgeoning network of day schools. Demand for places had grown so high that a second college, Southlands, was established across the Thames in Battersea in 1872, specifically for the training of women. Thereafter, Westminster functioned exclusively as a male institution, marking a decisive cultural shift in both character and ethos.[3]

Bell was 18 when he left the quiet certainties of his hometown for the great sprawl of the English capital. A letter home, written days after arriving in London, captures all the excitement and eager anticipation of a young man experiencing independence for the first time:

> 'As I told you last night I arrived safe and sound. Temperton turned up at Sheffield and turned out to be a right champion fellow. We had some fun when we got to London. We found the tube and arrived at Charing Cross all serene then our troubles began. The buses never seemed to stop and we were both heavy laden and were not acrobats. At last we decided to take a taxi which only cost us a bob… We have been through Westminster Abbey today and in St James' Park. This afternoon we had holiday and went into the Art Gallery and in fact all over the place. You can tell we saw something when we were on the move for three hours. There were some suffragettes outside the House of Commons delivering pamphlets. We hung about a bit to see if anything was going to happen, but we had no luck. There was a young girl, so we got her grinning. Barton who is a comical joker said she had disgraced the league. She was after a voter not a vote.'[4]

The same letter also carried news on the football front. 'I have just had a letter from the secretary of Crystal Palace,' Don revealed with obvious pride. 'He asked me to make an appointment with him.'[5] The secretary in question was Edmund Goodman, who had taken on the role of secretary and manager following the departure of Jack Robson in 1907.[6] A tough Birmingham native, Goodman had seen his own playing career cut short by serious injury playing for Aston Villa reserves. So severe was the injury that it led to the amputation of his leg and nearly cost him his life. After leaving Villa, Goodman moved south to assist Palace during their formal establishment as a professional club in 1905 and took full control of team affairs two years later.[7] He stayed at the helm for eighteen years and still holds the distinction of being the longest-serving manager in the club's history.

It remains uncertain whether Bell ever made the appointment with Goodman or, indeed, any other club representative. Nonetheless, it is clear that contact between the two parties had already been established before his move south. On 23 July 1909, the *Swindon Advertiser* reported:

'The Crystal Palace Club have signed on Donald Bell, a left back from the Starbeck Club, near Harrogate. Bell, who is only 18 years of age, stands 6ft in height and weighs 13½ st. Two years ago he played for Leeds City in a North-Eastern League game at West Stanley, and afterwards signed a Midland League form. He was not, however, called upon to play. He will play as an amateur for the Crystal Palace.'[8]

Don may well have agreed to play for the South London club, but there is no evidence to suggest he made any appearances for them, competitive or otherwise. The *Beckenham Journal* did report that he was one of several amateurs 'to be given a trial during August'.[9] Yet he did not feature in their first trial match on 18 August, and the second, scheduled to be played three days later, does not appear to have taken place. Don himself made no further reference to Palace during his time in London, which suggests his connection with the club was fleeting.

The college day was as long as it was demanding, which many newcomers found overwhelming. Students rose each day at 6.30 am, apart from Sundays, when they were granted forty-five minutes longer in bed. After washing and dressing, they had to strip their beds and open their windows, before hurriedly dashing to the first lesson of the day, which began at 7.00 am and lasted fifty minutes. Then it was back to their rooms to quickly make their beds before heading to the dining hall for breakfast. Dinner was served at 1.55 pm, tea at 5.20 pm and supper at 8.45 pm. Classes often extended into the evening, sometimes as late as 8.30 pm, while Saturday mornings were also spent in the classroom.[10]

Don appears to have been reasonably content with the meals provided by the college, at least during these early days. Food was a frequent topic of conversation in his letters, especially in correspondence with his mother:

'The meals up to now have been alright. They gave us tea which consisted of bread and butter. The funniest fact about it is that you have to cut your own bread. There is a big round loaf and you tackle it in turn. The butter is put on the table in big lumps and you take as much as you like and put on as much as you like. At supper they gave us an egg each also there was cheese. We had quite a good breakfast. First porridge then bacon and finish up with bread and butter. Of course you help yourself to all these. Dinner consisted

> of cold shoulder of mutton and plum pie. So you see that we are not doing so very bad at all. That plum pie was beautiful.'[11]

Like all educational institutions, Westminster operated under a set of 'pettifogging restrictions',[12] and Don had to adhere to a multifarious set of rules and regulations. These were outlined in a welcome letter from the principal, Reverend H.B. Workman, sent to him before he arrived. It discussed such matters as fees, vaccinations and accommodation, while also offering guidance on clothing – including the need for 'three weeks' supply of underclothing', a 'special cap' for weekdays and a 'frock coat and silk hat for Sundays'.[13]

The importance of communal worship had been emphasized since the foundation of the college, and students were expected to 'attend the Wesleyan Church, Westminster, on the morning and evening of every Lord's day, and shall, in the matter of religious observance, conform to Methodist rules and usage'.[14] During the week, he had to be in his room by 10.00 pm, with his lights out twenty minutes later. 'We have funny little bedrooms,' he told his mother, 'about 10ft by 8ft.'[15] On Saturdays, the curfew was slightly more relaxed, though students still had to be in their rooms by 10.30 pm, with all lights off by 11.00 pm. Dormitories remained off-limits during the day, except for half an hour following tea and on Sunday afternoons. Prayers were held each morning at 8.50 am in the lecture hall and at 8.00 am on Sundays in the dining hall. Evening prayers, meanwhile, took place immediately after supper, also in the dining hall.[16]

Students followed a curriculum that varied according to their chosen degree, though all programmes were rigorous. Don's timetable included English Language, Literature and Composition, History and Geography, Elementary Mathematics (including Arithmetic), Elementary Science (comprising Chemistry, Physics and Botany), Latin and French. Other subjects were compulsory for all students. These included the Principles and Practice of Teaching, Theory of Music and Singing, Reading and Recitation, and Biblical Literature. In addition, several hours each week were spent in the gymnasium, cultivating the physical strength, endurance and resilience that embodied 'Muscular Christianity'.[17]

Teaching practice also formed a core component of the weekly schedule. Previously, it had taken place exclusively within the college's own training schools. However, in 1908, Millbank School was used for the first time, with two students assigned to two-week placements in each of its departments. This change eased pressure on the college's own facilities and enabled students to complete their teaching practice in schools across the wider London area.[18]

Given the exacting demands of their timetable, it is remarkable that students found any time for recreational pursuits. Yet Don's letters reveal that an inordinate amount of sport was played during his two years at Westminster. Indeed, it is sport that drives much of the narrative of his time in the capital. This continued to bring him into conflict with his father, and it is revealing that correspondence between the pair was far less regular than with his mother and sisters. The few letters that do survive suggest Don often took a more pragmatic approach to maintain good relations:

> 'You do not write often but your letters always contain something pleasant. You need not be afraid about the work, pa, because I never mention it. We get plenty done, I can tell you. I just tell you the special things that happen, the things which vary the ordinary course. The tuck box has just come at the right time ie. for our weekly Sunday feed. The first course will be pork pie. The football team had a grand time yesterday. We were playing Islington Day Training College & beat them on their own ground 2-1. We went to the college for tea & had a splendid time. After tea we had a smoking concert of which I am enclosing a programme… I have not begun smoking yet & I am no nearer starting yet although I am chaffed frequently about it. You see I spend most of my time in the Smokers Room playing ping pong at which I am really quite an expert.'[19]

Don found a natural outlet for his sporting aspirations at Westminster, although the college had been slow to establish its credentials in this area. Hindered by limited space and inadequate facilities, little emphasis had previously been placed on athletic pursuits, and its record of success in most games was described as 'not impressive'.[20] At the turn of the century, chess, billiards and fives attracted chief attention – largely because they could be played on the premises. Change only began when the principal, determined that the college should excel beyond the academic sphere, secured the services of a man who would go on to have a significant influence on both Westminster's and Don's sporting development.[21]

Leigh Smith, known affectionately as 'Smiggy', arrived at the college in 1905 as a house tutor and lecturer in English History and the Classics, but his influence extended far beyond the classroom. With his 'boundless and contagious enthusiasm',[22] the Lancastrian played a central role in the development of the college's athletic programme and was instrumental in transforming its fortunes on the sports field. A graduate of Durham University,[23] Smith had previously taught at Harrogate College before moving south and was likely already known to Don in some capacity. An effervescent character and accomplished athlete in his own right, Smith had represented the university in several sports and played Minor Counties cricket for Durham.[24] His arrival marked a turning point for the college and also for Don, who would come to regard him as both a friend and a mentor.

Yet Smith's task had not been an easy one. Westminster had a deplorable athletics record and had endured 'disastrous results' at its debut inter-college sports meeting in 1904.[25] Nevertheless, he approached the challenge with verve and determination. Regular evening training runs were organized, mainly through the streets surrounding the Gas Light and Coke Company and Vincent Square, while plans were drawn up for the college's first inter-year athletics meeting. These efforts came to fruition on a crisp February afternoon two years later, when the seniors faced off against the juniors in a series of track and field events at the Tufnell Park athletics ground. Smith had persuaded the principal that holding the meeting at a recognized venue would have a positive psychological effect on the participants, and so it proved. The meeting was a roaring success and enthusiastically received by both competitors and spectators, although one observer would later reflect:

> 'How much the promise of a release from private study on the night of the sports had to do with their popularity it would be impertinent to inquire, but despite the frosty hand of the weather that day, the sports were held, the Seniors winning by a comfortable margin. Seven years were to elapse before, in 1912, the Juniors beat the Seniors.'[26]

Smith also encouraged students to compete in events and meetings further afield, and many did. The benefits were soon evident. At the 1907 inter-college sports meeting, Westminster made significant gains, finishing a creditable third in the points table.[27] This marked improvement was reflected in other sports too, further bolstering their emerging reputation.

The college's first great sports team was built around a nucleus of talented athletes, but none more prominent than Harold Ward, who arrived at Horseferry Road in 1906. A prodigious jumper, hurdler and sprinter, Ward was the quintessential all-round athlete and set all manner of records during his time at the college.[28] His greatest moment came at the 1908 inter-college athletics meeting, when he captained Westminster to its first overall victory.[29] It was a landmark day for the college: Ward and teammate Herbert Chesterfield shared first place in the high jump, while Alf Grey and William Singleton set new records in the quarter-mile and one-mile events, respectively. It was the second consecutive year that Ward had triumphed in the high jump, having cleared a record height of 5ft 3in twelve months earlier.[30]

Ward was undoubtedly one of Westminster's earliest bona fide sports heroes, setting a precedent that many would strive to emulate, including Don himself. However, it was Leigh Smith who emerged as the driving force behind the remarkable 'resuscitation of sport at Westminster'.[31] In 1907, the rugby XV took to the field for the first time and only lost once all season. The formation of a hockey club followed shortly thereafter. By the time Don arrived in 1909, the college was regularly fielding two competitive teams in football, rugby, hockey and cricket, while swimming and tennis remained incredibly popular.

Concerted efforts had also begun to find a more suitable and more permanent sports ground, prompted by continued uncertainty surrounding the existing pitch at Battersea Park. After a brief tenure at Wormwood Scrubs, the college finally secured a reliable venue at Boston Road, near Brentford. By 1909, the new ground had become home, with one or more of the college's teams playing there each Tuesday, Wednesday and Saturday afternoon.[32] It marked a remarkable transformation from the lacklustre environment Smith had encountered just four years earlier, progress from which all sports-minded students would reap the rewards.

Don first represented Westminster within weeks of starting at the college, scoring three tries in a one-sided 'rugger' match against Park House. 'I carried four of them over with me,' he bragged to his mother. 'It was laughable in the line outs to see four men marking me and leaving the others unmarked, consequently they were able to get through and we won 34 points to nil.'[33] Other appearances were less productive, including the annual seniors versus juniors match, when he was 'too well watched to get through'.[34] Even so, Don had evidently made a strong impression. In a letter to his sister Dolly later that month, he boldly reported that 'Smiggy' had said, 'if I kept on at Rugger, I was sure of an International cap', adding: 'What do you think of that?'[35]

While Bell's accomplishments on the football and rugby fields naturally draw most attention, they represent only part of his burgeoning reputation. During his time in London, Don pulled on the college colours in a range of other sports, including cricket, hockey and swimming, and served as sports captain in his final year. So great was his influence that he was later acclaimed as 'one of the finest all-round athletes Westminster has ever produced'.[36]

Equally noteworthy were his sporting activities beyond college circles. In addition to playing football in Harrogate during the holidays, often alongside his brother Billy, he regularly appeared for London amateur side Barnet[37] and also turned out for Hertfordshire in the Southern Counties Amateur Championship. After returning from one match on the south coast against Sussex, he wrote a rare letter to his father:

> 'We had a champion time down at Eastbourne. It was a lovely day we played on a lovely ground and the game was a good one. I was on my best form for once and the Capt told me after the game that I had played a "topping" game. It said in the paper we were lucky to draw. To tell the truth Sussex only had most of the play at the beginning and the end. As for their missing goals well, they had not time to steady themselves we saw to that. We had a bon tea and then went on the promenade and the pier till train time. The only blow on the day was that I had to pay part of my own expenses.'[38]

On 15 February 1910, Bell took part in his first inter-year athletics meeting at Tufnell Park. The event, and the raucous celebrations that traditionally followed, had become a staple of the collegiate calendar and drew enthusiastic participation. Don competed in several disciplines and performed strongly against stiff competition to finish second in the quarter-mile final and third in the 100 yards, shot put and high jump.[39] He may have been disappointed not to win at least one event, but it was an impressive day's work, nonetheless. It also secured his place in the Westminster team for the annual inter-college athletics meeting at Stamford Bridge on 5 May.

Featuring London's five metropolitan residential colleges and governed by Amateur Athletic Association rules, the meeting saw Westminster compete for the coveted Challenge Shield against teams from St John's College, Battersea; St Mark's College, Chelsea (the holders); St Mary's College, Hammersmith; and Borough Road College, Isleworth.[40] It was usually a keenly contested event and a popular fixture in the student year. However, with only two places available per team per event, competition was fierce, and Don was only selected for the penultimate event of the day, the quarter-mile.

It proved to be an exceptionally fast race, won by Charlie Russell of St Mark's in a record-equalling time of fifty-four seconds, despite dire conditions on the track. Don finished outside the top three, missing a place on the podium, but the experience was invaluable. Any personal disappointment was also likely offset by the announcement that Westminster had edged out the holders, St Mark's, to secure first place in the final standings:

> 'It would be a difficult matter to find a better sporting gathering than that of the Metropolitan Residential Colleges, whose inter-collegiate sports were decided yesterday afternoon at Stamford Bridge... The day was not a favourable one for fast performances, a strong wind interfering with the runners and jumpers, while the track was rather dead, owing to the rain, and the sprint and hurdles were run against the wind... A feature of the meeting was the enthusiasm displayed by the students from the different colleges, who attended in force, and with their various "cries" and flag waving brought to mind the American contingent at the Olympic Games a couple of years ago.'[41]

The team's exploits at Stamford Bridge were the subject of much chatter at the college the next day, with results reportedly 'broadcast with zest – and occasionally melody'.[42] However, the jubilant mood was soured later that day by news that King Edward VII had died at Buckingham Palace.

The 68-year-old monarch's death was not entirely unexpected; he was a habitual smoker and had been in ill health for some time, but his demise had a considerable impact. More than 250,000 mourners are estimated to have filed past his oak coffin as it lay in state for three days at Westminster Hall, with at least 25,000 turned away on the first day alone. Public buildings were draped in black, flags flew at half-mast and memorial events were held in cities and towns throughout the country,[43] including at Westminster College, where a service took place in the lecture hall on 12 May. Proceedings began with a solemn rendition of the national anthem, followed by the reading of psalms and the singing of hymns. The principal then addressed the students before the singing of the hymn 'Days and moments quickly fly', and the benediction brought the service to a close.[44]

On the day of the state funeral, private study and afternoon lectures were cancelled, giving staff and students the opportunity to watch the king's coffin being drawn on a gun carriage from Buckingham Palace to Westminster Hall, where a short ceremony was held. The cortege then moved down Whitehall and The Mall, passed Hyde Park Corner and Marble Arch, and continued to Paddington Station, where the coffin was transferred by train to Windsor Castle for burial.[45] Don was among the crowds lining the streets that day, though mention of it is curiously absent from his surviving letters. A year later, as the coronation of King George V approached, he confided to his mother: 'I don't intend going to see the procession, I had enough at the King's funeral and this will be 10 times as bad because for one thing the route is much shorter.'[46]

Beyond academic study and sport, music occupied a prominent place in the life of Westminster College, with a number of concerts held annually. The programme for the 1911 annual combined concert reveals that Don sang Tchaikovsky's 'Don Juan's Serenade' between a pianoforte duet, two dances from 'Nell Gwynn' by Yendall and Kenett, and a recitation of 'Jud Browning's account of Rubenstein's playing' by Singleton.[47] Students also performed concerts at other venues across London. These events were generally very popular and well-received, though Don did admit to his mother that they were often accompanied by some rowdiness:

> 'We gave a concert at Hanwell last Saturday and had a ripping time. I sang "Betty's Way". I wish you had sent Paddies Perplexity those two would have done champion. Hanwell is a long way out and we did not arrive home until 12.30. We visited a Fair at Hanwell and had some sport but the best fun was in the street in front of the big butchers shops. They had men outside

yelling away so we used to line up outside and have combines such as "Going 4d–3d–2d – send it down to the college for a penny", and them we struck up "The good old College Motto". The butchers quite enjoyed it because we drew a crowd. I went down to Mile End on Sunday and sang Two Old Stages.'[48]

By 1911, Bell was being sent to Herne Hill with teammate Fred Carrack to train under highly regarded polytechnic athletics coach Fred Clark. The arrangement represented a significant opportunity, though it did not come without cost. Thankfully, Westminster agreed to cover both the subscription fees and travel expenses of the two men, who were undertaking their teaching practice at the same school, as Bell related in a letter to his sister Minnie:

'Well, Smiggy has put Carrack and myself in special training under the Polytechnic, and we go to Herne Hill three times a week. We train in "Classy" company I can tell you. On Friday night we were running with Higgs who does the 100yds in 10 and one 5th of a second and Nicol the Polytechnic 440yds champion whose time is 50 3/5 so you see we are running with some class. I have had to give up football because it interferes with running so I shall not be doing any when I come home. I don't see very well how I can, seeing that the coll are paying our subscriptions to the Polytechnic amounting to 12/6 and every time we go down to Herne Hill our expenses are about 3/- so that you see I ought to give something up when so much is being done for us.'[49]

On 14 February 1911, Westminster held its sixth inter-year sports meeting at Tufnell Park. As sports captain, Don carried additional responsibilities, which he outlined in a letter to his sister Gertie: 'I have to make a speech and give a palaver on Tuesday.' Still, he acknowledged his role came with certain benefits: 'I am certain of a 1st Medal this time because the sports Capt gets one given besides what he wins.'[50] In the end, he had little cause for concern – he bagged five medals in total, including first place in the 440 yards. The following day, he wrote another letter to Gertie:

'Tell Pa that at last I have been able to get something in the sports getting 5 medals in all. The track was very sloppy which prevented any very good times being done. As you will see by the programme, I was 2nd to Carrack in the 100yds being beaten by a yard. I got 2nd in the shot doing 29ft 5½ inches to Sellers 30ft 3½ inches. My best performance was in the high jump where I

> was again second clearing 5' 2" and Farnham clearing 5'3½ inches which is a ½ inch better than either Inter Yr or Inter Coll record. It was a jolly good performance seeing that the track was sodden. My last event was the 440yds which I won. I did not go the pace as we are playing Boro Rd today and I did not wish to spoil myself. You will notice that all the distance events are done in poor times due to the state of the ground. There was a Polytechnic man down yesterday who wishes Carrack and myself to go in special training under Clarke of Herne Hill who is a very good man. I had to specify at the end and thank Mrs Prinny for distributing the medals. As you can see by the programme, we gave the Juniors a jolly good licking beating them by 31pts to 8 and as I got 9 pts I beat them on my own. Table V got 15pts out of 28 so we did very well.'[51]

Attention now turned to the inter-college meeting at Stamford Bridge, scheduled for 13 May 1911. As defending champions, Westminster were intent on retaining the title. 'We are very busy now training for the sports and I have got plenty to do,' Don told Gertie. 'We started the training tables last Thursday and are doing very well… Smiggy thinks we have a very good chance for the shield again and you can depend upon it we shall go all the way.'[52]

Leigh Smith's confidence was well-founded: Westminster romped to a convincing victory on the day, finishing seven points ahead of second-placed St Mary's. Don won his heat in the 100 yards but was narrowly beaten in the final by his friend and Herne Hill training partner, Fred Carrack. The *Sporting Life* reported that Carrack 'was best away, and was leading by a foot at 20yds, which advantage he maintained to the end, despite a fine effort by Bell'.[53] Don also had to settle for second place in the 440 yards, falling just short of victory in a closely fought race.

As sports captain, it was undoubtedly among Bell's proudest moments at Westminster, particularly as he was presented with the Challenge Shield by Lord Desborough, himself a distinguished athlete in his youth. The success prompted the principal to note in his logbook: 'Inter-College Sports at Stamford Bridge. A very fine day & good sport was enjoyed. Westminster easily won the Challenge Shield by 18 ¾ points against the next highest 11/10 of St. Mary's.'[54] Two days later, a report of the event also appeared in *The Sportsman*:

> 'Ideal weather was associated with the annual sports of the Metropolitan Residential Colleges for Schoolmasters at Stamford Bridge on Saturday and the 2300 spectators present witnessed some keen contests, in which previous records – besides being twice beaten – were also twice equalled. The former honours fell to C.E. Russell, St. Mark's, and R. Farnham, Westminster, in the Quarter-mile and High Jump respectively. The latter, but for stumbling at the last hurdle when leading, might have secured another record in the 120 Yards Hurdle Race, the winner, S. Church, Borough-road, just getting up to equal the previous best. F. Carrick also equalled record in the Sprint, but probably the most exciting finish of the afternoon was put in by J.C. Connolly, St. Mary's, who, running with good judgement in the One Mile Race, after a rare finish, just won on the tape. The successful competitors were received with intense enthusiasm by the respective Colleges, and Westminster College had good cause to be elated, again annexing the Challenge Shield with 18 3-10 points, St. Mary's being a good second. This was presented by Lord Desborough at the close of the meeting. His lordship, who had closely followed the events, gave a well-deserved compliment to both the competitors and executive on the success of the meeting. The medals were handed to the successful contestants by Mr Norman F. Hallows, the old Blue and Olympic champion.'[55]

Don's studies were now drawing to a close. On 6 June 1911, the seniors held a farewell concert for friends and family in the college lecture hall.[56] This annual occasion, which served as both a celebration of their time at Westminster and a gesture of appreciation to the staff who had guided them, was traditionally a poignant occasion. A more spirited farewell supper followed in the dining hall at the end of the month. Don's two-year association with the college came to an end when it officially closed for the summer on 3 July 1911, after which he returned to Harrogate, ready for the next chapter of his life.

Chapter 3

A Man of Two Callings

The *Daily Mail* Circuit of Britain Air Race was hailed as 'the greatest aerial contest ever held'[1] by local newspapers, and on Monday, 24 July 1911, it brought Harrogate to a standstill. An estimated 70,000 spectators are said to have paid between sixpence and five shillings to secure a vantage point on the Stray, while another 100,000 'found their spots free of charge'.[2] The unprecedented number of motor vehicles that converged on the town blocked roads in all directions, while at the station, a steady stream of trains ferried in people from across northern England throughout the day. Crowds stretched as far as the eye could see, as thousands waited with eager anticipation to watch 'some of the finest descents and ascents ever made in connection with aviation'.[3]

As it transpired, only five of the seventeen aircraft that took off from Hendon Aerodrome early that morning successfully navigated the 182-mile flight north, and even then, some rode their luck. One aviator 'just missed the top of the Royal Hotel as he came in low over the trees',[4] while another appears to have had a narrow escape after landing unconscious, 'slumped in his seat, his hands clenched around his steering wheel'.[5] Still, after a quick break for refuelling and refreshments, all five machines were soon back in the air and on their way to Newcastle, where they took another scheduled stop before pressing on to Edinburgh to complete the stage by dusk.

The event generated considerable public interest, and most of the town's residents, particularly its children, were desperate to witness the action. With the midsummer holidays also just around the corner, it is perhaps unsurprising that attendance at Starbeck Council School on the day of the race was recorded as being 'exceedingly low'.[6] In fact, by the time Lieutenant Jean Louis Conneau, flying under the pseudonym André Beaumont, brought his Blériot XI monoplane into Brooklands Aerodrome on 5 August to claim the £10,000 prize, the school had already been closed for a week.

When the school reopened its doors on 31 August 1911, it welcomed several new members of staff. Among them was Donald Bell, recently returned from his studies in London, who now took up his first post as an assistant schoolmaster. It was certainly a popular appointment. Don had made a significant impression at the school during his time as a student-teacher, and everyone was delighted to welcome him back. While he had been aware of the upcoming vacancy for some time, there had been some uncertainty in the months prior to his appointment. 'I have not heard from Butterworth yet about the Starbeck post,' he had complained to sister Gertie after submitting his application at the start of the year. 'I do hope Miss Pickard sent in her resignation before the meeting or I shall not know until June.'[7] With no further mention of the matter in his Westminster letters, it can be reasonably assumed that Miss Pickard had indeed submitted her letter of notice.

There were other staff departures during this period. Notably, Miss Daisy Thoseby, who left to take up a position at Grove Road Junior School. The 23-year-old was the daughter of the late Reverend William Thoseby and half-sister of Albert Edward Thoseby, headmaster of the Municipal Secondary Day School and the man credited with overseeing its transformation from a collection of rented rooms on Haywra Crescent to the large and prosperous grammar school it would later become.[8]

Unfortunately, the official record of Bell's time teaching at Starbeck is fragmented. He was initially placed in charge of Miss Pickard's former class, teaching pupils who had attained Standard VI, but was promptly reassigned to a class at Standard VII after a reshuffle by the headmaster. Beyond that, source material is notably lacking. No personal correspondence from the period is known to exist, and the school logbooks that cover the early months of his tenure end abruptly on Shrove Tuesday, 1912. Subsequent references are sporadic, confined largely to routine entries noting weather extremes such as blizzards and heatwaves, or outbreaks of illness, including scarlet fever, diphtheria, chickenpox and mumps.[9]

In the absence of evidence documenting Bell's life in the classroom, the narrative drifts once more to sport, in particular his flourishing football career. Two years older and physically stronger, Don was developing as a player and had already begun to establish a fine reputation for himself. A string of impressive performances in the local amateur leagues soon drew the attention of several professional clubs in the region, including one of the most prestigious.

Newcastle United were, at that time, widely regarded as one of the best sides in the country, having enjoyed spectacular recent success under the tutelage of Frank Watt. In little more than six years, the club had secured three league titles and reached the FA Cup final five times, though they were beaten on four of those occasions. They were also looking to bolster their squad.[10]

It is unclear when Bell first came to the attention of United, but he was invited up to Tyneside ahead of the 1911/12 season and was no doubt elated to be given the opportunity to prove his worth with such a prominent club. His first chance came on Wednesday, 16 August 1911, when he took part in the first of the club's three annual trial matches at St James' Park.[11] These fixtures, commonplace at the time, were used to assess players and raise money for local charities. Featuring a side made up of first-team regulars, the 'stripes', against a side of reserves and hopefuls, the 'whites', Bell was on the losing side, but gave a good account of himself. He impressed again three days later when the 'stripes' beat the 'whites' 2–1 in front of 12,000 spectators. His performance prompted the *Sheffield Daily Telegraph* to note: 'An amateur right back – Donald Bell – pleased the crowd.'[12] The final trial match, and the last chance to impress the club directors, took place on 26 August, when the 'stripes' romped to a 4–2 win.[13]

Bell's displays were good enough to persuade United that he had potential, and at the end of the month, he signed amateur forms with the club. It was a significant opportunity for the 20-year-old, but one tempered by the knowledge that playing time would almost certainly be restricted. Ahead of him in the pecking order were players of considerable renown and stature, including Billy McCracken, club captain and the man widely credited with the development of the offside trap; Frank Hudspeth, a future captain and club stalwart;[14] and Tony Whitson, an uncompromising South African enforcer. Any immediate opportunities would most likely come with the reserve side; however, competition there was equally fierce, with the likes of Bobby Waugh, Charlie Betts and promising teenager Dick Little all vying for selection, though Betts would soon depart for Derby County.[15]

Beyond the North-East, Bell's move garnered little attention. The signing of a relatively unknown young amateur was understandably overshadowed by the high-profile acquisitions of Billy Hibbert, signed from Bury for a record fee of £1,950,[16] and Scotland captain Jimmy Hay, who joined from Celtic for £1,250.[17] Among the few local newspapers to mention Don by name was the *Football Gazette and Telegraph*, which noted: 'The new men signed on by the St. James' Park club have shown excellent form. The Harrogate amateur, Donald Bell, is a big, strapping fellow, and a good player to boot.'[18] The *Newcastle Chronicle*, meanwhile, also made brief mention of the 'promising youth'.[19]

Bell made his reserve team debut on 30 September 1911 in a North-Eastern League fixture against Carlisle United at a sparsely populated St James' Park. The game was a one-sided affair, with the home side dominant throughout in a 4–1 victory. Had it not been for some wasteful finishing and good goalkeeping, one correspondent noted, the score could have 'run into double figures'.[20] Bell was paired at the back with the experienced Jack Carr, entering his fourteenth season at the club, and put in an assured performance next to the 33-year-old.

As an amateur, Bell was not contractually bound to one club. Two weeks later, he was back in the North-East to play for Bishop Auckland in an English Amateur Cup first round tie against Spennymoor at Kingsway. Around 6,000 fans packed into the ground, and the majority went home satisfied as the hosts cantered into the second round with a comfortable 3–1 victory.[21] Despite making a second reserve team appearance for Newcastle United in a 5–1 rout of Sunderland Rovers on 21 October,[22] Bell found playing time hard to come by over the next few months. He was hampered both by logistics and the fine form of others. Bobby Waugh and Frank Hudspeth had both started the season in imperious fashion, quickly establishing themselves as the first-choice defensive pairing. Waugh's outstanding form was recognized in October when he was one of five United players chosen to represent the North-Eastern League in a game against the South-Eastern League at White Hart Lane, home of Tottenham Hotspur.[23]

Though opportunities at Newcastle remained limited, Bell continued to turn out for several amateur teams, including Bishop Auckland, Harrogate YMCA and Yorkshire Combination side Mirfield United. He was also part of a Harrogate League representative side that defeated a Bradford District XI 2–1 in a charity match at the County Ground on 11 November.[24] A week later, he returned to Newcastle for his third, and as it would prove final, reserve team appearance of the year, contributing to a 1–0 victory over Spen Black and White in the Newcastle Infirmary Cup.[25]

On Saturday, 18 November 1911, Bell starred in Harrogate YMCA's emphatic 6–1 Whitworth Cup win against Ripon Celtic at Kirkby Road. Despite the one-sided nature of the scoreline, reports noted that the members of both teams had a 'thoroughly enjoyable day',[26] with tea provided after the match by the hosts and the visitors reciprocating with a short concert. Before returning home later that afternoon, Bell and his teammates were treated to a tour of the town hall by Councillor Savill.[27]

Away from football, life went on. Later that month, Don's eldest sister, Nellie, gave birth to her second child, Donald Bell Jackson, at the home she shared with her husband, Harry, on Silverfields Road, near the family laundry. Don was no doubt delighted to learn his nephew had been named after him, though perhaps not entirely surprised given that the couple had already named their first child after his elder brother, Billy. He adored his nephews and developed a close and tender relationship with them. The strength of this bond is reflected in Don's wartime correspondence, which make frequent reference to both boys.

In the decades following his uncle's death, Donald Bell Jackson would forge a fine reputation of his own on the football field, demonstrating that he had inherited more than just a name. A gifted schoolboy player, he went on to captain the student team at Manchester University, where he too trained to become a teacher, and also played for Northern Nomads, Yorkshire Amateurs and Leeds United 'A'. The pinnacle of his career came in 1935, when he led the Great Britain football team at the International University Games in Budapest, also acting as flag bearer. After graduating, he returned to Harrogate and taught at Grove Road Council School.[28]

Starbeck School closed its doors for the Christmas holidays on 22 December, allowing Bell more time for his sporting commitments. Just over a week later, the *Yorkshire Evening Post* reported that he had been invited on Bishop Auckland's upcoming overseas tour: 'Donald Bell, who formerly played with the Starbeck team, and now assists Newcastle and Bishop

Auckland, has been invited to accompany the latter club on its Easter tour to Venice, Vienna, and Buda Pesth [*sic*].'[29] While the prospect of playing on the continent must have been appealing, he declined the offer due to his teaching obligations.[30]

A congested fixture schedule over the festive period saw Newcastle's first team play five times in ten days, and the second string five in eleven. This included consecutive matches on Christmas Day and Boxing Day, which was customary at the time. This gruelling run of games inevitably took its toll, and by mid-January, the club had racked up a 'big list of players on sick leave'.[31] With so many absentees, Bell appeared set for another run out with the reserves. On 5 January 1912, it was reported that he would partner club stalwart, and current trainer, Andy McCombie at the back for a game against North Shields at St James' Park the next day.[32] McCombie, a former Scottish international who had won three English league titles with United and one with bitter rivals Sunderland, had been coaxed out of retirement to provide cover but would retain his place for the remainder of the season. However, for reasons unknown, it was Jack Carr who eventually took to the field with McCombie as United dispatched their beleaguered opponents by eight goals to nil.[33]

Bell's wait for his next opportunity went on, though fortunately, not for much longer. On 20 January, he played in a 5–0 victory at Seaham Harbour, and would surely have been optimistic about securing further game time in the weeks ahead.[34] The second string had scored a remarkable twenty-eight goals in five games, conceding just two, and sat second in the table, eight points behind Middlesbrough Reserves but with five games in hand. Yet Bell would make just one further appearance for the club, playing in the 1–0 reverse at Darlington on 24 February 1912.[35] Thereafter, his playing career, if not his association, with Newcastle United appears to have come to an end.

Bell's unexplained absence from the team sheet for the remainder of the season did not go unnoticed. The *Football Gazette and Telegraph* correspondent 'Felix' asked: 'What has become of Donald Bell, the amateur back at Newcastle United, thought so much of at the beginning of the season? Surely, he is an improvement upon a certain regular player at present?'[36] Why he was omitted from the team sheet remains a mystery. Injury was evidently not a factor, as he continued to appear for local sides during this period. The most plausible explanation is that teaching responsibilities took precedence. Don was fast approaching the completion of his first year and would soon be a fully certified schoolmaster.

On Saturday, 6 April 1912, the day after he had been scheduled to represent the Leeds Amateur League in a match against a Northern League side at Willington, Bell played for Harrogate YMCA in the final of the Ripon Charity Cup at Clotherholme Road. The opposition was provided by local rivals and reigning champions West Park. There was 'a large and enthusiastic gathering'[37] in attendance, with both clubs bringing 'strong contingents of supporters who displayed the colours of their respective clubs'. West Park added to their 'encouraging cheers' with the 'continuous clanging of a big bell'.[38] Strong winds made shooting particularly difficult during a hard-fought but goalless first half. The second half continued in a similar pattern, with few chances created by either side. However, with ninety minutes almost up and the tie still deadlocked, West Park managed to carve out one last chance:

> 'Just on time Forsyth ran out to check another attack, but the forward play was too strong, and after a scramble in the goal mouth Beal rolled the ball into the undefended net amidst the ringing cheers of the cupholders' supporters.'[39]

After being presented with the cup in the pavilion following the match, the West Park captain noted that 'it had been a good and fair game', adding that 'he hoped the Y.M.C.A. would take their defeat in a sportsmanlike manner, which he believed they would'.[40] No response from the losing team was recorded, so we will never know. In any case, Don had little time to dwell on the result. Just three days later, he lined up for Mirfield United in the final of the Bradford Hospital Cup against Bradford City Reserves. Nearly 3,000 spectators turned up to watch the match, played at Halifax Town's Sandhall ground, and they witnessed a fairly even first half. Unfortunately, things were less competitive after the break and the side from Bradford eventually romped to a 4–1 victory to secure the cup.[41]

Bell had one further opportunity to lift some silverware on 13 April, when Mirfield travelled to Dewsbury to contest the final of the West Riding Cup with Halifax Town. Mirfield were the reigning cup holders, having thrashed Allerton Bywater 5–1 twelve months earlier, and were overwhelming favourites. Though they 'took some time to find their feet',[42] once they did, the game became largely one-sided, even if the final score did not reflect their dominance. A host of good opportunities were squandered before they finally took the lead, and several more went begging before they added a second. According to one reporter, Halifax were 'never in the hunt', with the *Halifax Evening Courier* correspondent, writing under the name 'Pioneer,' concluding ruefully:

> 'It would be well if a veil could be drawn over the proceeding at Savile Town, Dewsbury, on Saturday, the scene of the West Riding Cup final in which Halifax Town took part. But hard facts and a disagreeable task must be faced. Mirfield United secured the cup for the second season in succession by two goals to none, a result that is far from giving an accurate reflex of the play. Had the score been doubled or trebled it would be more in harmony with the run of the game. Halifax Town were rendered impotent or did render themselves impotent?'[43]

In contrast to their defeat against Bradford City three days earlier, Mirfield were clearly the superior side. While several players delivered notable performances, Bell was the pick of the bunch. His dominance was such that 'Pioneer' observed: 'Culpan [the Halifax outside left] was a pawn in the hands of Bell.'[44] *The Leeds Mercury* agreed that he was the best player on the park:

> 'Mirfield throughout were the faster and more methodical side. Their forwards were in good line, whilst Moon and Horton in the middle division, rendered valuable assistance. Donald Bell, however, the Harrogate back, who assisted Mirfield, was generally admitted to be the best man on the field.'[45]

The following day, the same newspaper carried far graver news. At 11.40 pm on 14 April 1912, the much-heralded White Star Line luxury ocean liner, RMS *Titanic*, struck an iceberg off the coast of Newfoundland and went down with the loss of over 1,500 lives.[46] The sinking shocked the nation and would cast a long shadow over the remainder of the year, as inquiries and recriminations into the tragedy began.

When the 1911/12 Football League season finally drew to a conclusion on 27 April, Newcastle United found themselves in third place, five points behind champions Blackburn Rovers and two adrift of second-placed Everton.[47] It marked a disappointing end to the campaign, especially given that they had entered the new year four points clear at the top of the table and seemingly on course for a fourth league title. Injuries had taken a significant toll, a calamitous run of seven defeats in ten outings ultimately proving fatal. Their troubles were compounded by a dismal first-round exit in the FA Cup at the hands of Second Division side, Derby County, in January.[48]

In the North-Eastern League, the reserve side took the runners-up spot, three points behind Middlesbrough Reserves, with Darlington a further four points back in third. They also reached the final of the Newcastle Senior Cup, where they faced North Shields Athletic on 29 April.[49] The match ended goalless, and a replay was scheduled for the following day, with all gate receipts pledged to the Titanic Relief Fund. However, due to a shortage of available players, the match was eventually cancelled and the trophy shared between the two sides.

Bell was still on the books at Newcastle United at the start of the 1912/13 season, listed as one of seven amateurs, alongside namesake and local full-back George Bell, in a total squad of forty-one players.[50] He was named in the line-up for one of the 'whites' versus 'stripes' practice matches, but bad weather forced its postponement, and there is no evidence to suggest Don ever pulled on the club's colours again.[51] He continued to play locally, however, appearing for Harrogate YMCA in a 1–1 draw against Otley Old Boys on 14 September 1912. Following the match, the *Wharfedale & Airedale Observer* noted: 'The visitors also had a strong side out, including "Don" Bell, a local full-back much sought after in higher circles of football.'[52]

Bell's time at Newcastle United had now clearly come to an end. Yet, as one door closed, another opened, and he would seize the new opportunity with both hands.

Chapter 4

Up the Avenue!

Tom Maley arrived at English Second Division side Bradford Park Avenue in February 1911 with a point to prove. The enigmatic former Manchester City boss was among the most respected and influential figures in the game, but he had suffered an ignominious, albeit highly contentious, fall from grace in 1906 when a Football Association (FA) commission, initially set up to investigate allegations of match-fixing, found the Hyde Road club guilty of making illegal payments to players to circumvent the £4 maximum weekly wage.[1]

In a scandal that sent shockwaves through the sport, seventeen current and former players were banned for seven months, fined a combined sum of £900 and forbidden from ever pulling on a City shirt again. Several directors were also sanctioned, and the club itself fined £250. But the harshest penalties were reserved for Maley and former chairman Waltham Forrest, who were both banned from football *sine die*.[2]

The punishments were unprecedented and in the eyes of many, unjust. Maley vehemently denied any wrongdoing, arguing that the practice of paying bonuses was so widespread, in the top flight at least, that not four clubs 'would come out scathless [*sic*]' should the investigation be widened.[3] But it was to no avail. The FA stood firm and Maley, banished from the game he loved, returned to Scotland, where he would spend the next four years working as a school superintendent in the east end of Glasgow.

Then, in the summer of 1910, came a second chance. After rejecting several appeals to lift Maley's suspension, the FA finally acquiesced, granting him permission to resume his once-stellar career.[4] Still only 46, Maley was not short of suitors. He was even linked with a dramatic return to Hyde Road,[5] where his successor, Harry Newbould, had just guided City back to the top division after one season away. However, it was ambitious Bradford Park Avenue, backed by their wealthy chairman and benefactor Harry Briggs, who eventually secured Maley's services, bringing him to the club six months later to succeed former boss George Gillies.

The Bradford club offered Maley an ideal environment to restore his tarnished reputation, yet the challenges he inherited were markedly different from those he had encountered in Manchester. City had just been relegated, but they finished the season in strong form and possessed a team bristling with young talent. With Maley at the helm, they cantered to the Second Division title in his first season and won the FA Cup twelve months later – the first major honour for either Manchester club. In fact, City came agonizingly close to clinching a remarkable league and cup double that season, but eventually had to settle for the runners-up spot, three points behind champions Sheffield Wednesday. A year later, City came third, two points behind Newcastle United and one behind Everton.[6] Then it all came crashing down.

In Bradford, Maley found a club languishing in the lower reaches of the English second tier. Just three years after their election to the Football League, and four since their controversial decision to abandon Northern Union – modern-day rugby league – in favour of association football, Avenue were a club mired in mediocrity.[7] Maley set about transforming their fortunes, but progress would take time. His tenure began with a change of jerseys. The traditional club colours of red, amber and black, inherited from the old rugby days, were jettisoned in favour of green and white hoops, a nod to his beloved Glasgow Celtic. Recruitment followed as Maley, seeking to assemble a side capable of mounting a serious promotion challenge, brought in seasoned players like Celtic's Willie Kivlichan, Dundee's Bert Dainty and George Halley from Kilmarnock, to bolster a squad already featuring established names such as Tommy Little, Sandy Watson, Danny Munro and Bob Mason.[8]

In November 1912, Maley made one of his most important signings, persuading prolific centre forward Jimmy Smith to move north from Brighton & Hove Albion.[9] Smith was an immediate success and would go on to establish himself as one of the club's first great players, scoring fifty-eight goals in 100 appearances before the war brought his career to a tragic end. Progress towards promotion was steady rather than spectacular. However, under Maley's tutelage, Avenue soon garnered a reputation for their refined passing game and exciting wing play.

It was during this period of development that Donald Bell joined the club, signing the first professional contract of his career in October 1912. The acquisition of high-profile players like Kivlichan and Smith understandably grabbed the headlines, but the *Yorkshire Evening Post* was keen to report on the club's other signings too:

> The Bradford Club have secured the transfer of Donald F. [*sic*] Bell, an amateur full back from Newcastle United. He will play to-day in the Combination team. Bell stands over 6ft. and weighs well over 13 stones. He has the reputation of being a very fast and clever full back.'[10]

The *Football Gazette and Telegraph* also carried news of Bell's arrival, noting: 'He has a good turn of speed, having won a London Polytechnic open handicap off a short mark.'[11]

As anticipated, he went straight into the side for the Yorkshire Combination match at Fryston Colliery later that day, putting in a solid performance during a goalless draw.[12] This impressive debut was followed by several more assured displays over the next few months, as he quickly established himself as a regular with the reserve team. Reports from this period also suggest Bell was part of an 'International YMCA Football Team' that toured Denmark that winter. However, aside from claims that he garnered 'golden opinions',[13] little evidence can be found to substantiate the story.

Back in Bradford, Bell's performances with the second string had not gone unnoticed. On 16 April 1913, he was handed his first-team debut in a league match at home against Wolverhampton Wanderers. Coming off the back of a 5–0 drubbing at Hull City four days earlier,[14] he was drafted into the side to provide defensive cover for the injured Arthur Dixon and was one of two enforced changes. Their opponents, several places above them in the table but in poor recent form, were making their second trip to Park Avenue in as many months, having suffered a disappointing 3–0 FA Cup loss to Maley's side on 1 February.[15]

The weather leading up to the game had been atrocious, resulting in what the press described as 'a meagre attendance'.[16] Nevertheless, those supporters that did brave the elements were treated to 'a fine exhibition' by the home team, who breezed to an emphatic 5–1 victory:

> 'Bradford were easily the superior team and Bell made a very creditable first appearance at left back, and promises to be a very useful acquisition, whilst Watson also had a sterling game. Scott is getting more accustomed to the right wing, and his form is more in line with his usual wing play. Parker and Thompson played well without being brilliant. Howie was the pick of the forwards and Little a good second. The wingers were not prominent but Smith's control of the ball was at times good. Most of the attacking of Wolves' was due to Harrison and some smart combination by Needham and Brooks. The half-back line was weakened whilst both Collins and Garratly were weak under pressure at backs. Peers played well in goal.'[17]

After such a dazzling display of attacking football, most newspaper reports understandably focused on the Avenue forward line, particularly the contributions of Howie and Little, who both scored twice. However, Bell also received notable praise for his confident debut at the back, the general consensus being that he had acquitted himself well after 'doing good service with the Reserves for some time'.[18] The *Yorkshire Post* correspondent was among those most impressed, writing: 'Bell, from the second team, was tried at left back for the first time in League football, and he created a very good impression.'[19] Others agreed. The *Sporting Life* thought he had 'played a very sound game',[20] while the *Bradford Daily Telegraph* noted he was 'doing well for his first appearance in League football'.[21]

Although Bell had acquitted himself well, the match against Wolves would prove to be his only involvement with the first team that season. The vastly experienced Sam Blackham came into the side for a 5–0 thrashing of Glossop three days later,[22] while Billy McConnell, an Irish amateur who had made his debut against Burnley earlier that month, was given another run out in a narrow defeat at Clapton Orient on the final day of the season.[23] Still, Bell had made a favourable impression, and his brief cameo only served to enhance his growing reputation within the club. He had also placed himself firmly in contention for more regular first-team football and could look forward to playing a more active role in Avenue's promotion push when the new campaign began.

The day after the Wolves match, Harrogate YMCA returned to Clotherholme Road to face Ripon City in the final of the Ripon Charity Cup. Bell, who had played in the 1–0 defeat to West Park at the same stage the previous year, was unavailable on this occasion and watched from the sidelines as an interested observer. The family was still represented, however, with Don's brother, Billy, taking to the field as part of the Harrogate forward line. Unfortunately, he too would end up on the losing side as Ripon edged a tight game 1–0 to lift the 'handsome silver trophy'[24] and consign the YMCA to their second successive final defeat.

The end of the football season marked the start of the cricket season, and the Bell brothers were soon donning their whites to play regularly in the local leagues. Both boys were exceptional cricketers and fine all-rounders. Don's skills were never more on show than during a remarkable match at Harrogate on 26 July 1913, when he plundered most of his side's runs with the bat before tearing through the opposition's batting line-up with the ball, securing an impressive ten-wicket haul. The *Yorkshire Evening Post* reported:

> 'I have received a newspaper cutting giving particulars of a big bowling feat at Harrogate on Saturday last. Knaresborough II collapsed in their match with the Wesleyans, Donald Bell playing skittles with the wickets of the visitors and bringing off a sensational performance by taking the whole side for 17 runs, nine batsmen being clean bowled and the other caught and bowled by the Harrogatonian. Going in first the Wesleyans made 82, D. Bell having also the lion's share in this department, and scoring 34, including a couple of sixes, and 3 fours. Donald Bell is the footballer who had a trial with Newcastle United, and who has signed on for the Bradford Club for the coming season.'[25]

Four weeks later, Don was pulling his football boots back on for Avenue's first public practice match of the season – a comfortable 4–1 victory for the 'Whites' over the 'Blues' on the evening of 21 August. He was one of several reserve team players aiming to break into the first team, and clearly made an impression. The following day, the *Leeds Mercury* correspondent wrote: 'Quite a feature of the game was the sterling defence of Bell and Macrill, the backs of the Whites, both of whom played for Bradford Reserves last season, and are now engaged as professionals.'[26]

A second practice match took place on the afternoon of 27 August, giving players and hopefuls one final opportunity to showcase their talents in front of their manager and club officials. According to the *Bradford Daily Telegraph*, 'there was nothing particularly thrilling' about the encounter, though it did add the game showed Avenue had 'a good deal of talent to choose from and a nice blending of experienced players with young and eager enthusiasts in the game who should be more heard of later'.[27]

Despite this favourable press, Bell began the new season playing Yorkshire Combination football. However, on 1 October 1913, he was drafted into the first team for a midweek trip across town to face Bradford City in the second round of the West Riding Senior Cup at Valley Parade.[28] The away side, who were the current holders, had enjoyed a fantastic start to the new campaign, winning four of their opening five matches, and were brimming with confidence. Nonetheless, with local bragging rights at stake, 'both clubs were very strongly represented,'[29] and City went into the tie as marginal favourites.

Despite the 'Wool City Derby' still being in its infancy, around 10,000 spectators passed through the turnstiles to watch the highly anticipated encounter, and they were well entertained:

> 'The game opened with the Avenue the more aggressive side, and by neat combination they invaded the home territory, but Leavey shot well wide. The home right wing challenged Bell, the Avenue left back, who was forced to give a corner, and from a well-placed kick from Bond, Murray headed just a trifle wide of the post – a narrow escape for the Second Leaguers. Immediately afterwards the City goal had an even more lucky escape, for after Leavey had crossed the ball from the Avenue's extreme left over to Munro on the right, Smith was given an opening a few yards from goal. The Park Avenue centre, however, only tapped the ball in the direction of goal, and Mellors, by going to the ground effected a save at the second attempt. Several raids by the City wing men looked dangerous, but the visiting defence was steady and Drabble was only called upon to save a few shots, chiefly by Bond, Fox, and Murray. The Avenue forwards repeatedly were pulled up by the referee when in dangerous positions for being offside. Taken altogether the game had been evenly contested in the first half, and at the interval there was no score.

'On the resumption City took up the running, and with the Avenue rather off their game Drabble had some difficulty in keeping his goal intact. At the other end Little sent a good centre across the goal, but it was not taken advantage of, and then Bond and Bookman had splendid runs which ended in the ball just going wide of the post on each occasion. Then for some time the play became most uninteresting, neither side being able to work harmoniously. Ten minutes from the end, however, Bond raced down the City right wing. When near the touch line Drabble came out of his goal and tried to knock him off the ball. Bond, however, got in his centre, and Storey, who was standing a yard from goal, turned the ball into an untenanted goal. The game seemed won for the City, but a minute before the final whistle went, Munro dashed away, and sending in a good centre, Smith broke past the backs, and amid great excitement turned the ball into the net past Mellors, who had advanced to meet him.'[30]

Avenue had given a superb account of themselves, proving they could hold their own against one of the strongest sides in the country. Bell played particularly well and received glowing praise: 'Bradford folk must have been well satisfied with the display of Donald Bell,' wrote the *Bradford Daily Telegraph* the following day. 'He plays a typical amateur game, but plays it remarkably well.'[31] The *Park Avenue Journa*l also highlighted his performance, noting it was all the more impressive given he had picked up an injury during the opening exchanges:

'One of the heroes of the match at Valley Parade last Wednesday was Donald Bell, who was most successful as Sam Blackham's deputy. Although he received a nasty knock in the early stages of the game he kept the "elusive Dicky" in check, and more than once frustrated dangerous movements.'[32]

The 'elusive Dicky' was City's diminutive outside-right and England international Richard 'Dicky' Bond – at the time, one of the best-known players in the country. Bond would make over 300 senior appearances for the club, scoring more than seventy goals, which is testament to the level of Bell's performance that day. Unfortunately, optimism was tempered by reports suggesting his injury could be more severe than initially thought:

'The injury to Bell, who played so well for Bradford in the cup-tie against Bradford City yesterday, is rather more serious than was at first supposed. He

> sustained a bad kick in the first few minutes of the game, but he was keen to do his best for his side, and remained on the field, even though the injury was a big handicap. Blackham is much better to-day, and if the improvement continues he will turn out again at Glossop on Saturday. Should both he and Bell be unfit Mackrill will be called upon.'[33]

With such a quick turnaround between matches, it was inevitable that Bell would not recover in time for the 2–1 loss at Glossop three days later.[34] Nor was he part of the side that slumped to a disappointing 2–0 home defeat against Stockport County on 11 October.[35]

He was soon on the mend, however, and on 18 October, he was handed his first league appearance of the season as high-flying Fulham came to town. With first-choice left-back Sam Blackham sidelined by injury, Bell was brought into the team to play alongside Sandy Watson. Despite some unfamiliarity with the position, he acquitted himself well:

> 'Blackham was unfit and so Donald Bell was again called upon – this time at left back. The reversal of positions caused a little strangeness between Watson and himself at times, and once Watson got in the way of Mason. But the backs played a good game, and Mason was equal to all requirements in goal.'[36]

After a cagey first half, the home side took the lead just minutes after the restart when 'a pretty piece of combination between Kivlichan, Little, and Smith, produced a goal by the last-named'.[37] Avenue dominated the rest of the game but spurned a host of good chances, particularly Tommy Little, who 'should have made at least two more goals, but his marksmanship was at fault'. They were nearly made to pay at the death when the visitors almost snatched an equalizer. Thankfully, Pearce's effort was kept out by Mason and the game finished 1–0.[38]

Bell was composed and assured throughout, demonstrating once again that he was a reliable and highly capable understudy when called upon. Moreover, there was now the tantalizing prospect of an extended run in the senior side for the first time in his professional career. Indeed, the next few weeks would be the busiest of his entire time at the club.

After turning out for the reserves in a midweek loss to former club Mirfield United,[39] Bell travelled to Nottingham with the first team on 25 October to take on league leaders Notts County. With Blackham still unavailable, he kept his place at the back next to Watson, but the pair faced a mighty challenge against their free-scoring opponents.

Despite being heavy underdogs, Avenue delivered arguably their best performance of the season to edge a five-goal thriller, 3–2. The result was all the more noteworthy given County were previously unbeaten at Meadow Lane and 'making a strong effort to get back to the higher circle'.[40] But the away side were excellent value for their win. 'On the day's play Bradford were the better side,' observed one reporter, 'and their victory was the reward of perseverance and skilful play.'[41] Another noted: 'Conditions favoured good football, and Bradford were there to supply it.'[42] He added: 'It was a rousing game, for the "Lambs" didn't like being taken down, and it showed good training on the part of the Park Avenue men that they finished with great determination after a hard and fast game.'[43]

The *Leeds Mercury*'s correspondent was just as complimentary in his match report, published two days later:

> 'Bradford accomplished a splendid achievement by proving the first team to beat Notts County at Meadow-lane [*sic*]. The result caused much disappointment among the 14,000 spectators. The magnificent way in which Bradford battled for the points throughout a contest which was thrilling to a degree, showed that ability and stamina were their possessions. They had a goal through in six minutes. Little took advantage of a weak clearance on the part of Iremonger, and headed the ball back into the goal. Peart levelled the scores twenty minutes later, and this was the extent of the damage up to the interval. Bradford drew ahead again four minutes after the interval, when Smith improved upon a well-placed corner-kick taken by Leavey, but only two minutes elapsed before Peart, taking advantage of a cannon between Bell and Mason, who tried to cope with a spinning ball, made the score two all. As can well be imagined, the excitement was intense, and a strenuous fight was waged for the lead. This Bradford secured half an hour from the finish, when Kivlichan centred perfectly after beating West, and Leavey netted the winning goal.'[44]

Although still early in the season, the eye-catching result moved Avenue up to seventh, level on points with four other teams and just two behind new league leaders Bury.[45] It was an exceptional performance from the away side, who now hoped to kick on in their pursuit of promotion. On Bell's contribution, the *Leeds Mercury* noted he 'was a steady and energetic partner to Watson and gave a first-rate account of himself'.[46]

One week later, he made his third consecutive league appearance, and his fourth overall, as Avenue entertained Leicester Fosse. Both teams had picked up twelve points so far, though the home side had played one game fewer, and little separated them on the day. Roared on by their biggest crowd of the season, Avenue prevailed, securing a second successive 2–1 victory after another hard-fought and engaging contest.[47] Most observers agreed that they had been the superior side and fully deserving of both points.

Even so, the *Leicester Daily Post*'s correspondent noted ruefully: 'Bradford were only on terms at the interval because Clay, by a crude stroke of bad luck, headed through his own goal.'[48] He had further complaints:

> 'It was the penalty that gave Bradford the lead. The referee was prompt in giving his decision, and if he was sufficiently convinced of the justice of the award – as he ought to be in these matters – it was surely unnecessary to consult first one and then the other linesmen on the point. In this case the official's action could only strengthen any partisan view that the referee was not too certain about the matter, and whether that was the case or not those consultations with the lesser officials which so often determine the fate of one of the sides do not tend to strengthen the authority of the referee.'[49]

Still, he did concede the hosts were the better team overall, writing: 'There is no weakness anywhere in the Park Avenue front line, and their superiority was marked in the speed and decision with which they concentrated on an opening. No time was lost, and few mistakes were made.'[50]

Maley's men had now climbed up to fourth in the table, one point behind leaders Notts County, with only goal averages separating them from second-placed Woolwich Arsenal and third-placed Bury.[51] Their rich vein of form had seen them take maximum points from their last three matches, and Bell and his teammates were in fine fettle as they travelled to Wolverhampton Wanderers for their next game on 8 November. It must have helped that they were also unbeaten at Molineux, having won one and drawn four of their five visits to the ground since being admitted to the Football League in 1908.[52]

After playing out a goalless draw in the same fixture the previous season, the *Bradford Daily Telegraph* was optimistic that Avenue could go one better this time: 'If the weather holds good and the ground conditions are permitted to be favourable, Bradford may be relied upon to put up a good exhibition, and a reproduction of recent form should give them a good chance to improve upon last year's result… Their supporters are expecting a satisfactory ending to their journey.'[53]

Unfortunately, it was, by all accounts, an exceedingly poor game, with few meaningful chances created by either side. The *Yorkshire Post* lambasted the performances of both teams, complaining that the ball was 'too often in the air', while 'the passing, generally speaking, left much to be desired'.[54] The *Bradford Daily Telegraph* concurred, adding: 'The want of success on Bradford's part so far as goal-getting was concerned appeared to be again due to too much inside forward play.'[55] It took a 75th minute strike from the hosts' Sam Brooks to separate the sides, though it was noted the goal was scored 'from a position Bradford unavailingly claimed was an offside one'.[56]

The defeat was Bell's first with the senior side, having enjoyed four consecutive victories since making his Football League debut, against the same opposition, in April. Nonetheless, most reports suggest he had a good game, particularly given he was up against the 'smart footwork'[57] of Brooks, who spent the afternoon marauding down the Wolves' left wing.

The *Athletic News* were certainly impressed by his defensive work, noting that Avenue goalkeeper Bob Mason had in front of him 'two sterling backs in Bell and Watson, who kept their rivals so well at bay that most of the shots which came in were from a fairly lengthy range'.[58] Regrettably, the winning goal came on one of the rare occasions Brooks managed to escape Don's attentions, slipping past both him and Ted Garry in the area to slot the ball past the advancing Mason. There was no blame apportioned, however. Instead, the *Bradford Daily Telegraph* attributed the defeat to 'Bradford adopting wrong tactics in relying too much upon their inside forwards, to the neglect of their wings, both of whom played well when they had the chance'.[59]

Avenue dropped to seventh in the table after the loss, but it was a temporary setback. They returned to winning ways a week later, beating Hull City 2–0 at Park Avenue[60] before securing a narrow 2–1 victory against Barnsley at Oakwell on 22 November.[61] Indeed, Maley's side embarked on a five-match winning streak, climbing to second in the table, level on points with leaders Notts County.[62]

Bell played no part in this impressive run of results. Despite doing all that was asked of him, his spell in the first team came to an end when Blackham finally recovered from the injury that had kept him on the sidelines since early October. Put straight back into the side to face Hull, Blackham would play in all but one of Avenue's twenty-nine remaining matches, consolidating his position as the club's undisputed first-choice right-back.

Don likely felt some sense of injustice at losing his place in the side, but he was also a pragmatist. Blackham was an exceptional player and had made almost eighty senior appearances since arriving from Barrow in 1911.[63] He had also formed a fine defensive partnership with Sandy Watson, signed shortly after,[64] and both men were immensely popular with supporters. The *Bradford Daily Telegraph* was sympathetic to Bell's cause, acknowledging his recent contributions. However, it also recognized the practical reality: 'Blackham has recovered from his injury, and though Bell has been playing in capital form, it is not surprising that the directors should have decided to return to their old back.'[65]

The 1–0 defeat at Molineux on 8 November 1913 proved to be Bell's final appearance in the Football League. It was not the end of his time at Park Avenue, however. There was still Yorkshire Combination football to be played, and he turned out regularly for the reserve team over the next few months as they were eventually crowned champions, five points clear of his former side, Mirfield United.[66]

The first team's league campaign entered its final Saturday on a knife-edge, with three sides still possessing a mathematical chance of securing promotion to the First Division alongside runaway leaders Notts County.[67] Avenue occupied that position going into their home match with Blackpool, and knew victory would all-but guarantee promotion to the English top flight for the first time in their short history. Woolwich Arsenal, level on points in third but with an inferior goal average, faced a challenging trip to Glossop in the Peak District. Leeds City, two points further back in fourth but with a better goal average, were also in contention. However, they needed to beat Birmingham at Elland Road and hope that both of their rivals lost.[68]

More than 28,000 spectators packed into Park Avenue to watch the match, and they were forced to endure a 'very uneasy half an hour' as Blackpool played with 'a coolness and skill which was in marked contrast to the obviousness nervousness of the home men'.[69] Luck favoured the hosts, however, and after almost falling behind themselves, Jimmy Smith gave Avenue the lead following 'a bit of quick neat work' with McCandless and Leavey. The goal, scored completely against the run of play, appears to have sparked Avenue into life, and thereafter 'a complete change came over their game'.[70] Smith bagged a second soon after to double their lead, while Little and McCandless got the third and fourth after the break. Blackpool did manage to pull one goal back through Lane, but it proved nothing more than a consolation. The match ended 4–1, sparking jubilant scenes:

> 'After it was over there was great enthusiasm on the part of the crowd, for the news was early received that the Arsenal's success at Glossop had fallen short [of] the necessary number of goals to serve their purpose, and the crowd surged round the Bradford pavilion to cheer the players. There were repeated calls for Howie, the captain of the side, but David was too modest to say a word. Eventually, however, Mr. Maley, the manager of the club, came forward and thanked all their supporters for the recognition of the team's efforts, adding that now they were a First Division club Bradford would seek to carve out a future of which no follower of the club need be ashamed.'[71]

Champagne corks popped and wine flowed freely that evening as the club hosted a celebratory dinner at Bradford's Midland Hotel. The seven-course banquet was attended by 'a company of over fifty guests', including club officials, players and a number of their friends. Among the latter, it was noted coyly, were 'several ladies'.[72] A menu from the event, signed by an assortment of staff and players, confirms Bell was in attendance. So too was a correspondent from the *Bradford Daily Telegraph*, whose account of the evening's merriment appeared in Monday's edition of the newspaper:

> 'An excellent meal was enjoyed, after which there were a few speeches, and several telegrams of congratulations were received. Amongst one of the earliest to arrive was a warm message from the Bradford Northern club. Mr. Hardcastle, the chairman of the Huddersfield Town Club, sent a hearty greeting, there

> was another from the West Riding Association, one signed "Peter" (in whom was recognised the genial manager of the Bradford City club), and another sending the congratulations of the City players and signed by the good sport George Robinson.'[73]

Avenue's promotion to the top table of English football marked a significant milestone for the city of Bradford. For the first time ever, its two professional clubs would compete side by side in the country's highest division. It was a remarkable accomplishment, particularly for a city of fewer than 280,000 inhabitants, and placed it on equal footing with established footballing powerhouses like London, Manchester and Liverpool, which all had two sides in the top flight but were substantially larger in both size and population. Even Sheffield, the only other city able to boast such a distinction, had a population nearly double that of Bradford.[74]

Perhaps even more impressive, promotion had been achieved within six seasons of the club's election to the Football League, and just seven years since the so-called 'great betrayal', when Avenue turned its back on Northern Union to play association football.[75] For Harry Briggs, a Machiavellian figure to his detractors, the club's meteoric rise served as justification for the controversial decision to switch codes in 1907. It marked the culmination of seven years' hard work and steady progress and fully vindicated the decision to appoint Tom Maley as manager.[76]

Unfortunately, by the time players and staff began reporting back to the club to prepare for Avenue's debut season among England's elite, football was already becoming a secondary concern to many people. After months of sabre-rattling across Europe, Britain was at war with Germany, and professional football was about to enter one of the most turbulent seasons in its history.

Chapter 5

Controversy and Condemnation

British Prime Minister Herbert Asquith was keen to pursue a 'business as usual' approach to daily life at the start of the First World War.[1] As a result, most football clubs were going about their operations much as they had always done on the eve of the new season. There had been some disruption, notably at Manchester City, who temporarily lost Hyde Road to several hundred horses of the Army Remount Service,[2] and at Newcastle United, where St James' Park was requisitioned by the Territorials for a fortnight.[3] However, these inconveniences had largely been resolved and preparations for the new season were already well under way at clubs around the country.

While there had been no formal announcement made about the forthcoming season, senior representatives from the FA, notably chairman Charles Clegg and secretary Frederick Wall, publicly affirmed their position that the full league programme should proceed as scheduled. They argued it would help sustain civilian morale and provide opportunities for recruitment and fundraising. Responding to suggestions that all football should be suspended, Clegg remained resolute: 'Having regard to the great anxieties which all must feel during the continuance of war I think total suspension would be mischievous rather than good.'[4] It was a sentiment shared in the corridors of power at both the Football League and the Southern League.

Football was not the only high-profile sport pressing ahead with competition. Cricket's County Championship was still in progress, though only a handful of matches remained,[5] while the start of the Northern Union's twentieth anniversary season was just around the corner.[6] Dozens of horse racing meetings were also taking place at racecourses across the country each week. Yet it was professional football that bore the brunt of the condemnation, drawing widespread criticism from sections of the press and several prominent individuals.

Unfortunately, comparisons with one of the country's most prominent bastions of amateurism, the Rugby Football Union (RFU), did little to advance its cause. Indeed, the response taken by the two codes at the outbreak of war could not have been more pronounced. The RFU acted swiftly and decisively, suspending all competition and placing Twickenham at the disposal of the War Office.[7] It also encouraged players to enlist. 'The Rugby Union are glad to know that large numbers of their players have already volunteered for service,' a communiqué sent to clubs read. 'They express a hope that all Rugby players will join some force in their own town or country.'[8] This they did, in their droves. So emphatic was the response that the sport quickly became the yardstick by which all others would be measured.

Football had several high-profile critics, and they ensured their voices were heard. Among the most vociferous were W.G. Grace, the country's most venerated amateur sportsman, and Sir Arthur Conan Doyle, the cantankerous creator of Sherlock Holmes. Grace, who had also criticized the decision not to suspend the County Championship, considered it morally reprehensible that organized sport should continue at a time of national crisis and made repeated calls for an outright ban. Conan Doyle was equally scathing in his condemnation of men choosing sport over enlistment. 'If the cricketer had a straight eye,' he declared, 'let him look along the barrel of a rifle. If a footballer had strength of limb let them serve and march in the field of battle.'[9] He went further in November, sneering that 'men who had no knowledge of drill or of the rifle were no more useful than women or professional football players'.[10]

Another critic, East End social reformer and temperance advocate Frederick Charrington, published a series of one-penny pamphlets lambasting football. He declared it 'a national shame and disgrace to our country if we have our best athletes charging one another on the football field instead of charging the Germans on the battlefield'. Charrington believed the country's 'seven thousand able-bodied and highly-trained football Athletes'[11] should set a stirring example by joining the colours instead of playing sport.

As critics ramped up their campaign against professional football, the sacrifices being made by the country's legion of amateur players, typically men from upper and middle-class backgrounds – many of whom were army reservists or territorial soldiers – were often, and conveniently, overlooked.

Like the RFU, the Amateur Football Association (AFA) suspended competition shortly after the declaration of war. Amateur footballers flocked to recruitment offices during the opening weeks of the war, with some clubs losing a substantial proportion of their players within the space of a few days. Corinthians FC, perhaps the most renowned of all amateur clubs, are a striking example. Fourteen of their players were en route to South America when war broke out. Arriving in Buenos Aires ahead of a scheduled tour of Argentina, four of them – all army reservists – immediately disembarked and boarded the next available ship home. The rest of the party followed several days later, arriving in London on 30 August. By the end of the war, twenty-two Corinthians were dead, including three of those who had departed for the ill-fated tour. Many more were wounded and would never kick a ball again.[12]

Although most were loath to acknowledge it at the time, the economic and legal ramifications of suspending the season were a matter of grave concern for many within the game. A significant number of professional players were married men with young families; a demographic not yet being called on to volunteer. With no alternative means of income should football be halted and contracts cancelled, the uncertainty caused considerable anxiety and left many struggling to reconcile their patriotic duty with the need for financial stability.

On 22 August 1914, the day before the British Expeditionary Force (BEF) fought its first major action of the war at Mons, clubs across the country took part in a series of matches to raise money for the Prince of Wales's National Relief Fund.[13] Four days later, Bradford Park Avenue held the first of their annual public practice matches, raising £27 6s for the fund, with a further £64 donated following their second practice game on 29 August.[14] By the end of the month, football had raised more than £7,000 for the war effort, considerably more than any other sport. It was a remarkable effort, but it did little to placate its detractors. A week later, Field Marshal Lord Roberts of Kandahar, one of the British Army's most distinguished old soldiers, added his voice to the argument. During an inspection of the newly formed 10th (Service) Battalion, Royal Fusiliers (City of London Regiment) at the capital's Temple Gardens, the dogmatic general told them:

> 'How very different is your action to that of the men, who can still go on with their cricket and football, as if the very existence of our nation were not at stake… This is not the time to play games, wholesome as they are in the days of piping peace.'[15]

On 31 August 1914, the eve of the new season, committees from the Football League, the Southern League and the FA met at their respective London headquarters to discuss the increasingly difficult situation. For those calling for an immediate suspension, the outcome of these talks would prove disappointing. Both the Football League and the Southern League announced that the season would proceed, though they did recommend that clubs provide basic military drill and shooting practice for their players.[16] For its part, the FA argued that the War Office had not formally requested they suspend competition, although it did issue a statement encouraging clubs to release players and staff from contracts so they could enlist without fear of legal repercussions. It also urged clubs to place their grounds at the disposal of the War Office on non-match days and to allow prominent public figures to address spectators in an effort to stimulate recruitment.[17]

Some 24,000 supporters passed through the turnstiles at Park Avenue on Saturday, 5 September 1914, to watch Bradford make their First Division debut against reigning champions Blackburn Rovers.[18] The bumper crowd reflected the excitement generated by promotion and showed public appetite for the game remained largely undiminished. For a sizeable proportion of the population, life continued as normal, which for many meant a Saturday afternoon stood on the terraces.

Avenue delivered a spirited performance against their star-studded opponents but ultimately fell to a 2–1 defeat. Still, it was an encouraging showing from the division's newcomers, prompting one correspondent to predict: 'The same form maintained against other teams should see them make progress in the League.'[19] The report in the *Yorkshire Evening Post* concurred in part, but adopted a more cautionary tone:

> 'Where Bradford require strengthening the most is in the full back position. The management are alive to the deficiency and have been negotiating for at least one new man, but up to the time of writing no definite pronouncement

> on the subject had been forthcoming. The reserve full backs are Bell and Mackrill. The former, who is a school teacher in the Harrogate district, played several good games with the League team at the beginning of last season, and Mackrill is a Girlington lad who is coming along well.'[20]

The following day, cricket's County Championship was brought to a premature conclusion. With only a small number of fixtures remaining, the impact was negligible.[21] Nevertheless, the decision gave campaigners renewed hope that football would now be compelled to follow suit. There was to be no backtrack, however, and within a week, the opening round of league fixtures were complete.

The anti-football lobby were indignant, launching a series of scathing attacks in the press. Dr Thomas Fry, Deacon of Lincoln and a former public school headmaster, called for a ban on football coupons and gambling, restrictions to prevent spectators under the age of 40 from entering grounds and the immediate cancellation of player contracts to allow enlistment.[22] Frederick Charrington even sent a telegram to King George V urging him to withdraw his royal patronage from the FA.[23]

There were also concerns from within the game, including signs of discontent at some of the country's leading amateur clubs. Among those to voice their opposition were Bishop Auckland, one of Donald Bell's former teams. Their robustly worded response to news that they would be required to fulfil their forthcoming Northern League fixtures was published in the *Sunderland Daily Echo* on 2 September:

> 'That this committee learns with regret that the English F.A. have decided to continue playing football during the ensuing season, notwithstanding the serious crisis through which the country is now passing, and that we decided to go on with our ordinary fixtures under protest.'[24]

Soon after, and under mounting public pressure, the FA announced that it had placed itself at the full disposal of the War Office and began efforts to increase match-day recruitment.[25] Posters were displayed at grounds across the country, while prominent figures were invited to deliver rousing addresses to spectators. The FA also requested official sanction to continue the season but made it clear it was prepared to halt all competition at once should the government consider this the best course of action.[26]

It was becoming increasingly obvious that the sharp decline in attendances, and consequently in gate receipts, at most grounds was pushing several clubs towards financial ruin. In the Second Division, season ticket revenue had fallen from an average of £516 to £177, while total gate receipts were projected to drop from £326,017 during the 1913/14 season to £210,620 by the end of the current campaign. Warnings were issued that the second tier would soon collapse under these dire conditions. In response, the country's highest-paid players were urged to accept pay cuts in a spirit of 'brotherhood'.[27]

Those on the maximum wage, which had risen to £5 per week in 1910,[28] agreed to a fifteen per cent reduction, while players earning less than £3 took a five per cent cut. The savings, estimated to be around £300, were paid into a war relief fund to aid clubs in financial difficulty. Further contingency measures, introduced four weeks later, saw clubs agree to contribute 2.5 per cent of their gross gate receipts to the fund. Nevertheless, pressure continued to mount.[29]

On 4 September, Bell and his Bradford teammates met to discuss the situation. It was subsequently agreed that, in addition to contributing a percentage of their weekly wage to the war fund, 'all games in their club rooms should have a levy placed upon them and that the proceeds of this levy should be added to the weekly subscriptions'.[30] The subject of military training was also discussed at length, as the *Bradford Daily Telegraph* reported later that day:

> 'In order to equip themselves for eventualities they are also placing themselves under a competent instructor to receive tuition in military drill. It is quite likely that through the generosity of a local body the players will be able to take advantage of a fully equipped rifle range on certain days of the week. The meeting was purely a players' meeting, but it had the full approval of the directors.'[31]

Avenue were not the only club to commit to military instruction. Brighton & Hove Albion were the first to take such measures, organizing drill and shooting practice under the guidance of Irish forward Charlie Webb, a former soldier with the 2nd Essex Regiment.[32] Across town at Valley Parade, Bradford City's players were soon being drilled by popular centre forward Harry Walden, another old soldier.[33] Elsewhere, Clapton Orient established a rifle club next to their ground, where staff, players and supporters were taught how to march and handle rifles.[34] Officials at Crystal Palace, meanwhile, proposed that players from every club in the capital should be 'placed at the disposal on the War Office for two days each week for drilling at Hyde Park, Clapham Common, or other suitable place', and that they should also be 'released from football training to practice rifle shooting under instructors provided by the War Office'.[35]

Avenue manager Tom Maley came from a military background. He was the son of a regular soldier from County Clare, and had two sons already serving their country. Yet he spoke with remarkable candour on the subject of enlistment and the continuance of football. In the face of growing vilification and even hate mail denouncing the club's directors as 'enemies of the country', he repeatedly and robustly defended the position of the game he loved:

> 'What are those who are left at home to do? It is best that they should follow healthy recreations, and if it comes to such a pass that all able-bodied men will be required for defence purposes it will be found that footballers will be ready to answer any call made upon them by their country.'[36]

Maley's second-eldest son, Josie, had enlisted with the 1/9th (Glasgow Highland) Battalion, Highland Light Infantry shortly after the outbreak of war and was, at that time, undergoing training before being sent to France. His eldest son, Thomas, had joined the Royal Navy at the end of October and was serving aboard HMS *Cumberland*. Both were capable athletes, particularly Josie, who had occasionally turned out alongside Bell for Bradford's second string. Maley later recalled Don's response to the news that his son had enlisted:

> 'At the outbreak of war, he spoke strongly about those who could go, ought to, and didn't. When he learned that my sons had joined up he was mighty pleased: Josie Maley he knew well, they having played together in the second team. He had a great regard for him – a feeling which Josie reciprocated.'[37]

Josie Maley eventually sailed for France in January 1915 and soon rose to the rank of corporal. By May, he had been accepted for a temporary commission and was scheduled to return to Britain to attend an officer training course. Tragically, just days before he was due to leave, he was badly wounded at Festubert and died shortly afterwards in hospital at Béthune. He was 21 years old.[38]

Tom Maley was profoundly affected by the death of his son, and reportedly attempted to enlist himself, though he was recognized and turned away on account of his age. Thereafter, he devoted much of his time to recruitment efforts, speaking with passion and eloquence at meetings across the country. Despite his personal loss, he remained steadfast in his support of football, consistently arguing that a ban on the game would be a 'ghastly mistake'.[39]

Bell's first-team opportunities at Park Avenue were now scarce, but there was still plenty of reserve team football to be played. On the day the first team lost to Blackburn Rovers, their second string were beaten 3–0 at Doncaster Rovers in the Midland League. It was a disappointing result, particularly as they had thrashed Hull City Reserves 5–2 in their opening game of the season just three days earlier. Still, they made amends a week later, when Bell starred in a comfortable 3–1 home win against Grimsby Town.

On 2 October, it was announced that the Bradford Lord Mayor's Fund had so far raised £25,687 for the war effort. This included a sum of £11 13s 2d donated by Bell and his teammates, which the local press noted was the 'first instalment contributed by players from their wages'.[40] A second sum, totalling £11 19s, was donated four weeks later, with further contributions made over the coming months.

Relinquishing part of his football income, difficult though it must have been on his modest teaching salary, had now become a secondary concern. Driven by a growing sense of moral and patriotic duty, and no doubt stung by public attacks on the game, Don had decided to enlist and had already submitted a formal request to be released from his contractual obligations:

> 'I have given the subject very serious consideration and have now come to the conclusion that I am duty bound to join the ranks. Will you therefore kindly ask the Directors of Bradford Football Club to release me from my engagement.'[41]

Don's brother Billy had suffered no such hesitation, travelling to York to join the Army Service Corps (ASC) just three days after war was declared. Remarkably, with a dearth of skilled motor mechanics at the front, Billy was rushed across the English Channel to a mechanical transport (MT) company less than a week after signing his attestation papers.[42]

On 19 October, the *Leeds Mercury* reported that the directors at Park Avenue had agreed to Don's request to have his contract terminated: 'Donald Bell, one of the reserve full backs of the Bradford Football Club, has asked to be released from his engagement in order to join the army. His request has been granted, and the directors have expressed their appreciation of his actions.'[43]

Don had given serious consideration to applying for a commission, confident he possessed the attributes sought by the army in its officers. Yet, after carefully weighing up his options, and determined to reach the front as quickly as possible, he concluded that the fastest route to France lay through the ranks. With his decision made, he travelled to York where, on 28 October 1914, he enlisted as a private in the 9th (Service) Battalion, The Prince of Wales's Own (West Yorkshire Regiment).[44] His choice did not meet with everyone's approval. Upon hearing that his brother had joined up, Billy Bell told their mother: 'I was very disappointed when Don enlisted as a private. He ought to have a commission.'[45]

As Bell began the transition to life under King's Regulations, the financial strain and denigration of football intensified. There were even calls made in the Houses of Parliament for legislation to be introduced to halt the season with immediate effect.[46] The game suffered another significant setback in late November when it was announced that the popular press would stop publishing football-related content, restricting coverage to results in dedicated sports publications only.[47] Local newspapers did not follow suit. Match reports, news and updates on players, particularly those who had enlisted, continued to occupy column space in publications around the country. Many of these reached the front, maintaining a sense of connection with life at home, though soldiers were also witnessing the increasingly public war of words being waged in the press.

On 28 November, the same day Bradford beat Sheffield United 2–0 at Park Avenue, the *Yorkshire Evening Post* informed readers that Don had officially relinquished his position at Starbeck School: 'Donald Bell, the Harrogate and Bradford Association player, has given up his position under the Harrogate Education Committee to join Kitchener's Army, with whom he is undergoing training.'[48]

Soon after, the FA entered further discussions with the War Office. These talks resulted in the cancellation of international fixtures, though calls to suspend the FA Cup were rejected. A proposal was also made to form a dedicated 'football battalion', similar to the hugely successful 'pals' battalions raised during Kitchener's initial recruitment drive. Players could already be found serving in regiments and corps throughout the army; nevertheless the idea was broadly welcomed. The overwhelming success of Sir George McCrae's 16th Royal Scots, which contained a significant number of Scottish players among its ranks, was cited as an example of what could be achieved.[49]

With the position of English football becoming increasingly untenable, action was taken. On 8 December, representatives from each of London's eleven professional clubs met with officials from the Football League, the Southern League and the FA to determine their next course of action. The talks were chaired by William Joynson-Hicks MP, the man tasked with raising the 'football battalion' by the War Office. Also present was Captain Thomas Whiffen, the army's chief recruiting officer for London.[50]

These preliminary discussions proved productive. On 15 December, the 17th (Service) Battalion (1st Football), The Duke of Cambridge's Own (Middlesex Regiment), was officially raised during a rambunctious meeting at Fulham Town Hall. By the close of proceedings, thirty-five players had stepped forward to join the new battalion. Among them were six from Croydon Common, four from Brighton & Hove Albion and three each from Chelsea and Watford. The most remarkable contribution came from Clapton Orient, where no fewer than ten of their players joined up, including club captain Fred 'Spider' Parker and fan favourite William Jonas. Bradford City centre half, and former regular soldier, Frank Buckley was also among the first to volunteer. He was promptly handed a commission on account of his previous experience.[51]

By mid-January, it was reported that the battalion had attracted almost half the number of volunteers required to bring it up to full strength. By the end of March, the battalion was complete and had already undergone several weeks of training at White City. The vast majority of recruits hailed from London and the South-East. Although it was initially hoped that volunteers would come from across the country, the cost of transporting players back from the capital to fulfil fixtures placed a significant financial strain on many clubs. As a consequence, recruitment beyond these regions, particularly among professional players, proved underwhelming. This prompted a response from the battalion's commanding officer, Colonel C.F. Grantham. Writing to the management committee of the Football League, he complained that 'only 122 professionals have joined', adding: 'I understand that there are forty League clubs and twenty in the Southern League, with an average of some thirty players fit to join the Colours, namely 1,800. These figures speak for themselves.'[52] More players did eventually arrive from the provinces, including six from Grimsby Town. Nevertheless, the 17th Middlesex, and their sister battalion the 23rd Middlesex (2nd Football), raised by Joynson-Hicks in May 1915, would remain, in composition, a predominantly London-centric battalion throughout their existence.

On 24 April 1915, twelve hours before the first Allied troops stepped ashore at Gallipoli, almost 50,000 spectators watched Sheffield United take on Chelsea in the final of the FA Cup at Old Trafford. Dubbed the 'Khaki Final' on account of the large number of uniformed soldiers in attendance, the match was a one-sided affair, with the Yorkshire side winning by three goals to nil. After presenting the cup to United skipper and man-of-the-match George Utley, Lord Derby addressed the players: 'To both winners and losers, I offer my most hearty congratulations. You have played with one another and against one another for the cup; play with one another for England now.'[53]

Four days later, Bradford Park Avenue hosted neighbours Bradford City in the final game of the season. The home side cruised to a comfortable 3–0 victory, securing both bragging rights and a ninth-place finish, one point and two places above their rivals. At the top of the table, Everton beat Oldham Athletic to the title by a single point, after the latter suffered a surprise home defeat to struggling Liverpool.[54] It would be the last time competitive top flight football was played for more than four years.

Within months, both the Football League and the Southern League had announced the suspension of their full league programmes for the duration of the war. The FA followed suit, issuing new regulations that not only confirmed the suspension of the FA Cup and the Amateur Cup, but also prohibited the registration and payment of players.[55] Though temporary, these measures effectively marked a return to amateurism – something opponents of professionalism had been pursuing since its introduction more than three decades earlier.

The decision to suspend official competition brought one of the most contentious periods in the history of English football to a conclusion. However, it did not mark the end of organized football. Regional competitions were soon established to fill the void, providing much-needed distraction for munitions workers, soldiers home on leave and those recovering from wounds.

The Football League oversaw its own wartime competition, divided into a Lancashire Section and a Midland Section, each comprising fourteen clubs.[56] The London Combination was re-formed, with seven clubs from the Southern League and five from the Football League.[57] A third competition, the South-Western Combination, was also established, with five English clubs and two Welsh, though it was short-lived and discontinued after a single season.[58] To avoid disruption, fixtures were restricted to 'Saturday afternoons, and on early closing and other recognised holidays'.[59] With official player registrations suspended, individuals were free to 'guest' for other clubs – typically those nearest to their army camp or munitions factory.

With the season at an end, several of Bell's teammates followed his lead and joined the colours. Prolific inside forward Tommy Little enlisted with the ASC;[60] Irish international Jack McCandless entered the ranks of the Glasgow Highlanders;[61] and Jimmy Smith, the club's star centre forward, joined the Motor Machine Gun Service.[62] Over the next three years, others would follow suit, including Ernie Scattergood, George Jobey and Willie Kivlichan. The war deeply affected the club, and, like many others, it would not emerge entirely unscathed.

Perhaps the final word should go to Tom Maley, a man who remained staunch in his support for the game, despite the devastating loss of his son. Throughout the war, he consistently argued the case for football enthusiasts on the home front, particularly those engaged in essential war work in the factories. Maley saw no contradiction in supporting the war effort while also advocating for the continuation of football as a vital form of recreation and distraction. It was, in his view, a matter of principle, and one he articulated with conviction: 'Munition workers deserve a little outdoor recreation, and football will give it to them. What do the anti-football folk offer to these jaded workers if the game of football is to be taken from them?'[63]

Chapter 6

King's Regulations

By 16 November 1915, the 11th (Reserve) Battalion, Alexandra, Princess of Wales's Own (Yorkshire Regiment) had been at Rugeley Camp for almost four weeks. One of two vast training camps built on the windblown expanse of Cannock Chase in Staffordshire, each large enough to house 20,000 men at any given time,[1] it was a bleak and barren spot capable of grinding down the most hale and hearty of men. But Second Lieutenant Donald Simpson Bell was in comparatively good spirits as he stepped off parade that morning.[2] Determined to make the most of his remaining time at home before embarking for France, he had arranged to travel up to Wilmslow at the weekend to visit his fiancée. He was rather looking forward to it.

Don began his courtship with Rhoda Margaret Bonson while they were both students in London. The daughter of a joiner and carpenter from Kirkby Stephen in Westmorland and three years his senior, Rhoda was already a trained dressmaker but was now studying to become a Methodist schoolmistress at Southlands College. Although she did go on to teach – including taking up a new post shortly before her marriage – details of her career in the classroom remain limited. By the time of the 1911 Census, Rhoda had returned to dressmaking and was living in Wilmslow, lodging with her uncle, George Bonson, and his wife, Annie.[3]

Bell had held his temporary commission scarcely twelve weeks, yet he had been in uniform for over a year and was growing increasingly frustrated by the continued delay in crossing the Channel. Within days of signing his attestation papers at York in October 1914, he had found himself en route to Belton Park near Grantham, where the 9th (Service) Battalion, The Prince of Wales's Own (West Yorkshire Regiment) was already two months into its initial phase of training with the 11th (Northern) Division. Constructed on the sprawling Lincolnshire estate of Adelbert, 3rd Earl Brownlow, Belton

Camp could already accommodate more than 13,000 men and functioned like a self-contained town. Navigating its labyrinth of starched white bell tents and wooden huts could prove a challenging prospect for the uninitiated, and many a new arrival would lose their bearings to incur the wrath of incredulous sergeant majors. Yet it was at this vast camp that Bell had a chance encounter with an old school friend that would alter the trajectory of his life.[4]

He had not seen Archie White, now a subaltern with the 6th Yorkshire Regiment, for several years when their paths crossed late one afternoon. The pair arranged to meet that evening and spent several hours reminiscing about their days together in Harrogate before the conversation turned, inevitably, to Don's decision to enlist as a private rather than pursue a commission like his friend. He was a natural leader and well-suited to army life. However, while he had already impressed enough to be appointed lance corporal, it was clear to both men that he had the makings of an exceptional officer. How much persuasion was required remains unknown, but by the end of the evening, Don had been convinced to apply for a commission with his friend's battalion.

White was asked about that night at Belton Park shortly before his death in 1970, and said this:

> 'I found him, as a L/Cpl wearing a green tweed overcoat, in the 9th West Yorks… I got him over to the next row of bell tents, which were the 6th Yorks. Cd. Chapman had one look at him and recommended him for a commission on the spot.'[5]

It was not quite 'on the spot'. Bell did not submit his application until 10 May 1915 – four weeks after the 9th West Yorkshires had relocated to Witley Camp in Surrey. Nevertheless, thereafter, things had moved at pace. Official confirmation arrived from the War Office at the end of the month and the commission was formally announced in the *London Gazette* on 8 June 1915.[6] Within a week, Bell was sat on a train heading up to Tyneside to attend an officer training course at the Tynemouth School of Instruction, housed at the Bath Assembly Rooms.

Alas, little evidence survives from which to construct a narrative of his time at Tynemouth. A thorough scouring of the official records yields no more than brief references to the school, and no correspondence from this period survive in the family archives, though some must surely have been written. We do, however, know that his stay began in dramatic circumstances.

Unbeknown to the newly commissioned officers settling down for their first night at the Grand Hotel, their lodgings for the duration of the course, a German Zeppelin was on its way across the North Sea to raid the industrial areas along the River Tyne, passing perilously close to Tynemouth as it did so. After making landfall near Blyth just before 11.30 pm, the airship travelled south and dropped its first bombs over Wallsend and Hebburn. It then crossed the river to target the Jarrow shipyards, guided by the light of Palmers' blast furnaces and engineering works, which bore the brunt of the damage. Further bombs were dropped north of the Tyne before the Zeppelin ditched its remaining payload over South Shields and made good its escape. The raid left eighteen people dead and seventy-two injured. It also caused over £40,000 worth of damage.[7] The sound of exploding ordnance and anti-aircraft fire was so loud that it jolted most of the guests at the Grand out of their beds. However, Don would later confess to his mother that he had 'slept through the affair'.[8]

Like all officers in the British Army, Bell was required to purchase his own uniform and equipment, though he received an initial allowance of £50 to offset the cost.[9] He also had to buy his own sidearm, the procurement of which could prove a confusing prospect for any new officer. Fortunately, his resourceful brother, Billy, was on hand to assist. Shortly before Don left for Tynemouth, Billy wrote to their mother reporting that he had secured both a revolver and a fine pair of black 'Military Stereo 6x30' binoculars. Don would later have these binoculars – manufactured in the United States by the Bausch & Lomb Optical Co. of Rochester, New York – engraved with his name. Billy's use of a rudimentary form of 'code' in his letter proved remarkably effective in evading the scrutiny of the army censor:

> 'I want you to tell Don that I have bought those two articles that he was wanting, i.e. Eyblwe ervolvre [Webley Revolver] and some ldefi esassgl [field glasses] which are brand new & ought to save him 6 and 8 pounds expense. I hope he has not forgot our arrangement.'[10]

The transaction would take place face to face, as Billy returned to Harrogate for his first period of home leave at the end of July. It was only a fleeting visit, and he was soon on his way back to France, but he did have enough time to attend chapel, where he 'sang a solo at the evening service'.[11]

Tynemouth's school of instruction could accommodate around fifty junior officers at any given time and typically employed two or three instructors, often 'dug-outs' or those medically unfit for overseas service. Its four-week course consisted primarily of lectures, drill and field exercises, and the sight of junior officers being put through their paces in the grounds of nearby Prior's Park, just behind the Bath Assembly Rooms,[12] soon became a familiar one.

Considerable emphasis was also placed on etiquette and 'mess and manners',[13] for certain standards of behaviour were now expected of these new 'temporary gentlemen'. In preparing for their first appearance in the officers' mess, new subalterns were warned to expect 'some chaff and one or two practical jokes'. In such circumstances, it was recommended that they remain 'natural above all things, but not bumptious' and 'courteous but not servile'. They were also reminded to '*never* mention a lady's name' in mess, nor draw their sword, 'even when asked to'.[14]

In reality, the course was severely limited in its scope, equipping new officers with only the most rudimentary of leadership skills before packing them off to their reserve battalions ready for posting overseas.[15] Most of them would have to learn 'on the job', relying on the guidance of more experienced officers and the wide range of pamphlets, training manuals and instructional texts available for private purchase.

By the end of 1915, it was clear that the schools of instruction were failing to produce enough new officers to meet demand. The course was also wholly inadequate to prepare them for front-line service. Steps to address these shortcomings arrived in the form of the Officer Cadet Battalions (OCBs) in February 1916, just weeks after conscription had been ushered in by the passing of the Military Service Act.[16] Thereafter, temporary commissions were only granted to those who had successfully graduated from an OCB.[17] The duration of the course was also extended, to sixteen weeks, although some were shortened in response to the heavy losses on the Somme. Cadets would now be assessed continuously throughout their training, which meant underperforming candidates could be identified and rooted out before reaching the front.[18]

If Bell had anticipated swift passage to France upon leaving Tynemouth, he was to be thoroughly disappointed. Instead, he was sent to the regiment's reserve battalion, the 11th Yorkshires, at Hummersknott Camp in Darlington, where training continued. Frustratingly, he was still with them in October when they left County Durham and moved into hutments at Rugeley Camp.[19]

Several of Bell's Bradford Park Avenue teammates had already beaten him to the front, despite having joined up much later. Among them was free-scoring inside forward Tommy Little, who Billy Bell came up against during a match between their respective ASC teams:

> 'We had a lovely match last Sunday against a team from the 20th Supply Col. I met Tommy Little of Bradford & had a chat. They also had Kirkman of Sheffield Wed playing for them & though giving away a soft goal they won 2–1. We hope to beat them in the return match next Sunday.'[20]

Billy also revealed that he had been contemplating applying for a commission himself. 'I've often thought about it,' he told his mother, 'but it would be in the infantry. I will let you know definitely soon.'[21]

In late August, a 27-year-old Scottish footballer serving with the 1/8th Highland Light Infantry, Lance Corporal William Angus, received the Victoria Cross from the king during an investiture at Buckingham Palace. The Carluke native, who had briefly been on the books at Celtic, was recognized for 'most conspicuous bravery and devotion to duty' at Givenchy on 12 June 1915, when he 'sustained about 40 wounds from bombs'[22] while rescuing a wounded officer from no-man's land. As a result, Angus, who also lost his left eye, became the first professional footballer to be awarded his country's highest military decoration for gallantry.

The much-anticipated War Office wire with the order for France could come at any time. For Donald Bell, it arrived just after morning parade on that bitterly cold November day at Rugeley Camp. 'My word, what a rush and bustle I had,' he wrote in a letter home later that day. 'First I had to go over to the Yorks orderly room for instructions and railway warrant. The C.O. was in and he told me I had to set off at once, although the Leicesters were given until tonight.'[23]

With barely enough time to pack, Bell enlisted the help of one of the officers' servants – often referred to as a batman, a term derived from the old French *bât*, meaning 'packsaddle'.[24] Unfortunately, things did not go well. 'I left my man to pack my things, but I find now that he has left several necessary things and put in articles I didn't want,' he complained later. 'Above all, he has forgotten my cardigan.'[25] He also had to leave without his watch, though this was not the fault of the hapless servant, as Bell had taken it to a local repair shop himself and not picked it up.[26]

After a hurried farewell to his comrades, Bell finally left Rugeley Camp late that afternoon and was soon sat on a train rattling south towards London. Significant delays along the route slowed his journey, and it was almost 10.30 pm when his train finally pulled into Euston Station. With no onward travel available until the next day, Bell headed south towards the Thames, eventually booking a 'very reasonable' room for the night at the Faulkner's Hotel on Villiers Street.[27]

Sat in the shadow of Charing Cross Station, the hotel was less than a mile from his old college on Horseferry Road, and early next morning, he walked over in the hope of seeing his old sports mentor, Leigh Smith, before catching his midday train to Folkestone. Unfortunately, it proved a wasted journey: the premises had recently been requisitioned and were now serving as the headquarters of the Australian armed services. The college itself remained operational, but it had been relocated to the Wesleyan Theological College in Richmond, Surrey, some ten miles away.[28] Still, that was of minor inconvenience compared to the upheaval faced by the students of the establishment, who had been displaced 200 miles north to Didsbury College in Manchester.[29] With time at a premium, a trip to Richmond was out of the question. Disappointed, Bell returned to the station and boarded his twelve o'clock train. He would never see old 'Smiggy' again.

The bustling south-coast town of Folkestone had excellent rail links with London and could be reached in little more than ninety minutes before the war. However, it was now the main embarkation port for troops heading to France,[30] meaning there were significantly more trains on the tracks.[31] This inevitably caused congestion, leading to frequent and often lengthy delays. 'I left London at 12 on a horrible train,' Don would bemoan to his parents, 'which did not reach Folkestone until 4.15 – 60 miles at the most.' Fortunately, a subaltern from the Leicestershire Regiment provided congenial company, and the two men took tea together when they reached Folkestone. When they finally made it to the docks, the silhouette of a large troopship loomed over the quay. Fully expecting to clamber on board and steam off into the night, they were surprised to be offered a choice of sailings by the accommodating embarkation officer. 'I wanted to go by daylight,' Don explained later, 'so we put off the evil day and are staying here for the night.'[32]

With overnight accommodation now required, the pair made their way up to the Hotel Metropole. Perched at the western end of the Leas, Folkestone's famous clifftop promenade, the Metropole was a grand, 200-bedroom Victorian behemoth, described at its opening in 1897 as 'second only to the Brighton Metropole in the boldness of its conception outside London'.[33] A popular haunt of officers, it boasted two dining rooms, a library, ballroom and smoking, billiards and drawing rooms, all finished in oak, gold lacquer and gilt, marble and velvet.[34] The hotel was used as a hospital for wounded Belgian soldiers at the start of the war, but had reopened to guests by the end of 1914. It was requisitioned again in 1917, this time by the British Army, and thereafter used to accommodate members of the Women's Army Auxiliary Corps.[35]

Before heading downstairs to eat that evening, Don wrote home. It had been an eventful couple of days, and he had much to report:

> Dear Everybody
> You must think I am taking a very long time to get across the water but am not sailing until tomorrow and now perhaps I had better give you a connected account of my doings up to the present moment. Yesterday morning I came off parade about 1.30 to be greeted with the news that I was for France and was to report at Folkestone as soon as possible. There were brothers of the Yorks to go but as word had come through at 10 o'clock they had already cleared. We had been engaged on a scheme 4 miles from camp or otherwise I should have been sent for... I left camp about 4-15 and caught the train to Birmingham. Unfortunately, I just missed the 6.20 to Euston and had to wait until 7 the latter not arriving in town until 10. I put up at Faulkner's Hotel for a night, very reasonable and next door to Charing Cross. During the morning I paid a visit to Westminster but did not see the Dr who is at Richmond... I hope you will remember me every week with a parcel but will tell you my requirements

later. Have put a fair reserve in the bank for the purpose which should see us through. Now I hope you won't worry about me because I am not going to take any unnecessary risks. I am a "Nosey Parker" in most things but I shall not be too anxious to take a peep at the German Trenches. I do not feel in the slightest worried and slept like a top last night. It is surprising for I am generally in a blue funk before anything like this comes off – especially with family complaints thrown in. I feel very disappointed that Rhoda and I could not have the weekend at Wilmslow as we had arranged but we cannot grumble. Before I forget see that Nellie has 10/- for a present for little Donny.

Must close now as it is dinner time and I am ready for it.

With love to all

Yours

Don

P.S. Am enclosing Nancy's letter.[36]

The weather the next day was glorious. Folkestone was bathed in bright morning sunshine, casting a golden hue across the whitewashed facades of its tall seafront townhouses, while the French coast at Cap Gris-Nez, little more than twenty miles away, was clearly visible on the horizon. The scene was reminiscent of a late August Bank Holiday, were it not for the November chill in the air and the proliferation of Royal Navy vessels in the Channel.

After making his way back to the docks, Bell found a grey-painted troopship already waiting for him, smoke billowing from its huge funnels. There would be no hero's send off, however. Fifteen months of war had taken its toll on the local population, and few in the town retained any real appetite for pomp or ceremony.[37] It stood in stark contrast to the scenes of August 1914, when fervent crowds packed the streets to cheer off the first contingent of the BEF. Some soldiers found the apathy of the locals galling, particularly those heading to France for the first time. Others were more understanding of the town's war-weariness. 'I was not surprised, nor distressed at the absence of ceremony,' wrote a subaltern from the Dorsets, 'nor the complete indifference of the population of Folkestone, who for more than a year witnessed the daily passage of troops on leave and reinforcements to France and Flanders.'[38]

Once on board, life jackets were distributed and men jostled for a place to sit down. A moment of quiet reflection was typical as the ship manoeuvred its way out of the harbour and into the open waters of the Channel. Most men contemplated their future and wondered whether they would see home again. The crossing usually took around two hours – sometimes less – but it was rarely comfortable, even for those fortunate enough to secure a place on the upper deck. For those confined to the bowels of the ship, the experience could be wretched. Many troopships were filthy, and most suffered from poor ventilation due to overcrowding. In heavy seas, the stench of vomit, sweat and human waste clung to the stagnant air, seeping into every crevice. Conditions were marginally better in fine weather, but much depended on the age of the vessel and the time of year.

Then there was the menace that lurked beneath the waves. The Dover Strait had been heavily patrolled by the Royal Navy since the start of the war, and a system of minefields and submerged indicator nets, known collectively as the Dover Barrage, stretched all the way to the Belgian coast.[39] Nevertheless, German submarines continued to pose a significant threat to Allied shipping, even after the suspension of their campaign of unrestricted warfare at the beginning of September.

Just one week before Bell sailed for France, the British hospital ship HMHS *Anglia*, carrying 390 wounded officers and men back from the front, struck a mine dropped by a U-boat and sank within fifteen minutes. Nearby vessels rushed to the scene to assist, but more than 160 lives were lost, including a nursing sister, nine members of the Royal Army Medical Corps and 129 wounded men.[40]

The sinking of the *Anglia* provoked widespread shock and outrage in Britain. Yet it served as a stark reminder, if one were needed, that crossing the Channel carried inherent risks, even under the most benign of conditions. Sailing to Boulogne on a bright, sunny day in 1916, one junior officer noted: 'This might be a summer holiday, except that astern there's a deadly-looking little destroyer: our escort, I suppose. Mines, torpedoes – the sea is full of man's filth: there is enough to provide a remote possibility we might never reach France.'[41]

Boulogne-sur-Mer was one of the primary base ports used by the BEF in France. It could be a bewildering place and had undergone a dramatic transformation since the first British troops stepped ashore there fifteen months earlier:

'What a change has come over Boulogne since August 1914! It has been converted from a guest-house into a workshop. ... Boulogne was French, and full of French excitement and cordiality. To-day, as we silently drifted in like cargo, and like cargo lay waiting I know not what formalities before we could land, I felt that Boulogne had been Anglicised.'[42]

After disembarking, those officers and men not proceeding directly to their units were marched through the cobbled streets and up to a large transit camp overlooking the town, where they remained until moving on to their next destination. Accommodation at the camp, a mixture of sparse wooden huts and canvas tents, was basic, even by army standards, and few got much sleep before reveille roused them again at 5.30 am.[43] Bell spent his first night in France here, and most of the next day, before boarding a train that would take him down the coast to a place that needed little introduction.

Situated fifteen miles due south of Boulogne, where the English Channel swallows the River Canche, Étaples was home to a sprawling amalgam of numbered infantry base depots (IBDs), training grounds, convalescent camps and general hospitals. It was a bleak wilderness of wooden huts and tents, capable of housing over 100,000 men at any one time, and had already acquired a notorious reputation by the end of 1915. War poet Wilfred Owen, who spent time there in January 1917, described it as 'neither France nor England, but a kind of paddock where the beasts are kept a few days before the shambles'.[44] Another officer called it 'the last place on earth', adding that 'the atmosphere surrounding the place was rotten'.[45]

The IBD was, in essence, a large holding camp where officers and men received further training while they waited for their posting. It was a place marked by restlessness and simmering tension, where resentment often ran high due to the harsh regime and the arbitrary discipline meted out by the camp's permanent staff. Much of the men's ire was reserved for the instructors, nicknamed 'canaries' on account of their yellow armbands, who harried and harassed them at every turn on the notorious 'bullring', a place writer Edmund Blunden described as that 'thirsty, savage, interminable training-ground beyond the dunes.[46]

Bell arrived at Étaples late on 17 November and reported to No. 17 IBD, where he was assigned to a tent with five other subalterns. The temperature had dropped sharply, and his first night under canvas was a miserable one. The following day, he wrote to his mother to tell her about his first few days in France:

> Dear Mother
> I have had a day at our new quarters so can give you some idea of our position. This is a base where we wait until posted to the battalion and therefore it depends upon the needs of the battalions whether I shall remain here long or not. We arrived here about 11 pm and were put under canvas, 6 officers in a tent. It is intensely cold but fine – in fact it is the most severe cold we have had yet and it is difficult to keep warm. We start parades tomorrow which I expect will be the usual thing, bayonet fighting etc. We are compelled to get our meals in town as there is not sufficient accommodations up in the camp. French cooking is rather different from ours. I am getting used to it, but not their prices, however, and it is my opinion many will retire on the money made out of "Les Anglais".
>
> Have very little this time except the address Reply by return as length of stay is uncertain.
>
> Love to all, Your affectionate son Don[47]

After finally deciding to pursue a commission, Don had set his hopes on joining his old school friend, Archie White, in the 6th Yorkshires. Unfortunately, the army sent men where they were needed, not necessarily where they wanted to go. At the start of July, White's battalion sailed for Gallipoli, landing at Suva Bay four weeks later.[48] This forced Don to reassess his options. 'I will let you know as soon as possible,' he told his mother, adding: 'I hope it will be the 1/5th as I know a good many of the officers in that battalion.'[49]

Étaples was a forbidding place, with a daily routine that was both rigid and intentionally demanding. Reveille sounded at 5.30 am, followed by breakfast, with troops parading at 6.30 am. All ranks, including officers, were then marched to their designated 'bullrings', where they spent the day engaged in such activities as trench digging, bomb throwing, unarmed combat and bayonet practice.[50] Gas huts also had to be visited to ensure that their flannel gas hoods were in good working order, a particularly loathsome experience, while there were all manner of lectures to attend for officers newly arrived in

France. A subaltern from the 10th Yorkshires was among those put through his paces during this period:

> 'We are on our way to the Bull Ring: two hundred of us, officers who have not been to the Front and are therefore due for a course of intensive training till some battalion of our regiments shall require us. Here we are, slogging along under the command of a captain, back in the ranks again, carrying rifles… Later we climb up among the sand dunes on the other side of the road, and there practise firing rifle grenades and throwing those small egg-shaped cast-iron missiles known as Mills' bombs. Here too we learn more of the methods of gas attack and defence, and practise the art of shoving our heads quickly into the clammy flannel bags that are dignified by the name of P. H. helmets. We finish the morning's work by running obstacle races over a prepared course back on the arena.'[51]

It was long and arduous work, but most men would come to appreciate the importance of what they were taught, if not the conduct of the instructors. 'We practised the most deadly, the most efficient methods of killing, or avoiding being killed,' recalled one veteran. 'This tough finishing school was as near to the real thing as possible.'[52]

With its 'fierce and vindictive atmosphere', life at the IBD tested the will and sapped the soul. Yet Bell and his fellow officers undoubtedly had it easier than men from the ranks. Their time there was generally shorter – typically no more than a week – and they were afforded far greater freedom, which most made full use of. Étaples may have been described by one soldier as 'a dirty, smelly conglomeration of ramshackle houses and mean streets',[53] but it was only a short walk across the *Pont des Trois Arches* rail bridge and was home to a number of cafés and *estaminets* – all serving watery beer, cheap 'plonk' and that staple of the British soldier behind the line: egg and chips. Don's letters reveal he ate in at least one of these cafés, though as a committed member of the Independent Order of Rechabites, a religious temperance and friendly society,[54] he was at least saved the extortionate prices of the awful alcohol on offer.

Officers seeking a more refined culinary experience could also venture across the River Canche into the salubrious surroundings of Paris-Plage, which remained strictly off-limits to the troops. The bucolic charm of this upmarket seaside resort, a favoured destination of the French and English elite before the war, 'put the thought of camps and army routine a thousand miles away', according to one junior officer.[55]

The transient nature of life at Étaples often gave rise to a sense of aimlessness. 'There you were, doing parades which couldn't interest you because you knew that you were only doing them to fill in time,' recalled a subaltern from the Royal West Kents. 'For the rest you wandered about, knowing that at any hour of the day or night you might be ordered away.'[56] Some officers even feared they would be forgotten altogether, left stranded at the camp while the war carried on without them. In hindsight, many may well have wished they had been.

Despite its notorious reputation, not everything at Étaples was unpleasant. Scattered throughout the camp and its surroundings were a number of canteens and rest huts, where predominantly young female volunteers served tea, sandwiches and an abundance of smiles. Typically operated by religious organizations like the YMCA and the Church Army, these huts were immensely popular, offering a semblance of normality and a rare opportunity to interact with the opposite sex. One young soldier undoubtedly spoke for many when he told his parents: 'When we get inside the Y.M.C.A. hut, we feel as if we are home again.'[57]

In addition to food, drink and a range of other comforts and everyday essentials, the huts stocked a seemingly inexhaustible supply of writing paper. As a consequence, thousands of letters bearing the familiar inverted red triangle of the YMCA or the stamps of other voluntary organizations passed daily through the letterboxes of homes across Britain and the Dominions. Remarkably, by the end of the war, the YMCA alone had issued more than 900 million sheets of paper to troops free of charge. This helped maintain a vital link between men at the front and their families at home. Many of these letters remain preserved in family archives today, including several written by Don and Billy Bell.[58]

More than a million officers and men would pass through Étaples between June 1915 and September 1917.[59] The brutal regime and repressive conditions they encountered there left many with deep-seated resentment, particularly towards the instructors, who had subjected them to unrelenting 'illiterate, unanswerable personal abuse'.[60] When simmering tensions finally erupted into four days of unrest in September 1917, few experienced officers were surprised. It was, according to Charles Carrington – a subaltern with the 1/5th Royal Warwicks who later received the Military Cross – a reaction against 'acts of petty tyranny by tactless officers'.[61] Another officer observed: 'There is no doubt that the men had grievances. A base camp is utterly different from one's own battalion, where officers and men are known to each other… This, and the tension of waiting for orders, made base camp an unpleasant experience for men in transit.'[62]

Thankfully, Don's time at Étaples was relatively brief. On the evening of 23 November 1915, he wrote to his mother to inform her he had received news:

> Dear Mother
> I wrote to Rhoda earlier this evening giving her my new address but that is no use as on returning to the lines I found out I am to go up to the line early in the morning. I believe practically every officer who came here this last week are [*sic*] for it so there is something doing. My address will be:
> 11th Yorks attd 9th
> 69th Brigade
> 23rd Division
> B.E.F.
> We leave here 6 am tomorrow and it is now 9.30pm so must stop writing.
> Nothing else
> Love to all.
> Don[63]

Chapter 7

Nursery Sector

Despite its rather formidable-sounding name, Fort Rompu is a nondescript cluster of houses and farm buildings pitched off the road between the French villages of Erquinghem-Lys and Sailly-sur-la-Lys, south-west of Armentières. Archaeological evidence suggests some form of military fortification did once occupy the site, traces of which are visible on the Napoleonic cadastre of 1833. However, by the time the first British troops arrived in the area in October 1914, all physical remains of it had long since vanished from the landscape.[1]

When the opposing armies failed to outflank each other during what became known as the 'race to the sea', and the front eventually stabilized, Fort Rompu found itself almost two miles behind the line. It occupied a relatively quiet sector, primarily used to billet troops rotating into divisional reserve from the trenches around Bois-Grenier.[2] Even so, it did not escape unscathed. By the end of 1915, most buildings in the area bore signs of extensive artillery damage and were subjected to regular, if mostly desultory, shellfire. It was one of hundreds of similar villages scattered along the front, where troops usually spent no more than a few days before moving out again, either back into the line or further into the rear for rest, reorganization or training.

After five frustrating but largely uneventful days on the French coast, Bell finally left Étaples by train just before dawn on 24 November 1915 and set off to join his new battalion. His destination, the town of Armentières, lay some 60 miles inland, close to the sector where the 23rd Division, which the battalion formed part of, was currently holding the line. Today, the journey is a seamless and efficient one, completed in as little as two hours on the high-speed trains of the *Société nationale des chemins de fer français*. The route itself remains mostly unchanged too. Unfortunately, all similarities end there, for Bell's journey was neither short nor straightforward. It was mired by delays, diversions and unscheduled stops, with the train moving so slowly along certain lengths of track that men were able to jump down and walk alongside it to stretch their legs.

To compound his misery, Bell was forced to disembark at the village of Steenbecque and undertake the final leg of his truncated journey by bus. Consequently, Armentières was not reached until the early hours of 25 November. It was an ordeal he later chronicled in a rare letter to his father:

> 'To start with after landing at Boulogne I went down the coast to the base at Étaples. When we had orders to join our regiments we travelled back to Boulogne through Calais, St Omer & Hazebrouck where we changed. We had another short journey to Steenbecque where we took the night bus. That is the nearest place on our sector where you can come by rail. The bus took us to divisional headquarters at Croix du Bac, only a small village south of S[teenbecque]. We then went on through Erquinghem to Armentières where we stayed the night. That is the large town I told you of. Next day we found that the Battalion were billeted at Fort Rompu on the road between Erquinghem & Sailly.'[3]

Raised at Richmond in September 1914, the 9th (Service) Battalion, Alexandra, Princess of Wales's Own (Yorkshire Regiment) – colloquially known as the Green Howards – spent very little time in the county whose name it bore. Within days of its formation, the battalion had entrained for Frensham in Surrey, where it was joined by the 8th Yorkshire Regiment, 11th West Yorkshire Regiment and 10th West Riding Regiment to form 69th Brigade, part of 23rd Division. This colourful mass exodus of Yorkshiremen, laden with an eclectic mix of personal kit and impedimenta, made quite an impression on the correspondent from the *Huddersfield Daily Examiner*:

> 'The men began to arrive on Saturday from the depots of the Yorkshire Regiment at Leeds, Halifax, York, Pontefract, Richmond, and Sunderland. Two colliers carried pit lamps, several carried accordions, and one fondled a black kitten which was evidently regarded as a mascot.'[4]

The majority of these original volunteers were drawn from the industrial areas around Middlesbrough and the east Durham coalfields. Others came from the pit villages and ironstone mines scattered across the North Yorkshire

Moors, while some enlisted from as far afield as Tyneside and Wearside. A considerable proportion were older men, and many were married with young families.[5] Watching them arrive in Surrey, one onlooker noted they were 'a muscular lot of men' who, when 'licked into shape, should prove an excellent reinforcement to the allied troops'.[6]

The battalion spent the better part of the next year training at various camps across southern England before finally embarking for France on the evening of 26 August 1915. Thereafter, things moved quickly. They were placed under the guidance of the 1st Royal Scots and 2nd Cameron Highlanders for an introduction to trench warfare soon after their arrival.[7] By the end of September, they had moved into the line near Bois-Grenier to take over positions on their own for the first time.[8]

The battalion suffered its first losses to enemy fire during the second week of October, though these were relatively light and not considered exceptional. The regimental history refers to them as 'trifling wastage'[9] and notes they were soon made good by the 'small drafts of men that turned up from time to time'.[10]

The next few weeks were spent rotating in and out of the line and passed largely without incident. However, catastrophe was narrowly avoided on 4 November when a billet at Rue Marle, occupied by men from the battalion, was struck by shellfire. Although there were no fatalities, thirteen men were wounded and subsequently evacuated to No. 3 Casualty Clearing Station at Bailleul.[11]

The 9th Yorkshires had been at Fort Rompu almost twelve hours when Bell arrived from Armentières just after midday on 25 November. It was their first day in divisional reserve, having been withdrawn from the line east of Bois-Grenier the previous evening, and much of it was spent 'settling into and improving'[12] billets damaged during a recent spell of bad weather. Even then, Don would admit to his mother that the men still had a 'pretty rough time in some wooden huts with canvas roofs', adding somewhat sheepishly that 'we officers are quite comfortable'.[13]

The brigade war diary notes that 'Platoon, Company and Regimental Training',[14] took place on 26 November, as did 'Grenadier and Machine Gun Classes'.[15] There was also a voracious appetite for men to furnish fatigue and working parties, with 500 assigned to such duties on that day alone.

The workload was unrelenting and the cause of considerable grumbling among the ranks.

Most soldiers received some form of rudimentary bomb training before going into the line. Still, it was recognized that many lacked the 'temperament or the qualifications necessary to make a really efficient grenadier'.[16] Consequently, the training of specialist bombers, still officially referred to as grenadiers, became a matter of pressing concern. Initial instruction was ad hoc, with its quality largely dependent on the proficiency of the battalion's bombing officer, who may have only received a few days of training himself. Attitudes began to shift following the introduction of the No. 5 Mk I 'Mills' grenade in May 1915,[17] and a proliferation of specialist bombing schools had been established across Britain by the end of the year, and in even greater numbers in France and Flanders.[18] Similar schools were also set up at Gallipoli and in Egypt, and later in the war in Salonika, Mesopotamia and Palestine.

The 'ideal' bomber was to be selected from 'the very best, bravest and steadiest in an emergency',[19] with preference given to tall soldiers and those with the longest reach. Perhaps unsurprisingly, and significantly, given Bell's eventual rise to battalion bombing officer, it was found that sportsmen and those fond of outdoor pursuits often made the best candidates, being 'the easiest to train'.[20] Indeed, with his powerful physique, athleticism and daring attitude, Bell was the archetypal bomber.

Bomb schools taught every aspect of grenade warfare. This included practical knowledge of the mechanisms of standard grenade types and how to handle them, throwing techniques, offensive and defensive bombing tactics, and the importance of teamwork and coordination. Men also learnt how to use enemy grenades with proficiency, including the German *Stielhandgranate*.[21] Course duration varied. Battalion-level bomb schools offered only basic instruction, sometimes over just a few hours, while men selected for specialist training were sent on more intensive courses, often for several weeks, at grenade schools of instruction located behind the line.[22]

Bell would soon learn that courses, of which there were many, could become the bane of a new subaltern's life at his unit. Within twenty-four hours of arriving at Fort Rompu, he was detailed to attend a signalling course at III Corps Headquarters, then located at Château Motte-aux-Bois near Hazebrouck. He wrote to his sister Gertie later that day:

> 'Ma will be relieved to hear that I am not destined to visit the trenches for 3 weeks at least as I am detailed to attend a course of signalling... As the place is at least 30 miles from the firing line there should be no course for any anxiety. I leave the battalion on the 1st of December & as the exact duration of the course is not known I can practically ensure being there till Xmas.'[23]

Before Don had time to post the letter, the course was cancelled, meaning he was bound for the trenches after all. Still, it would not be his first time under fire. On 27 November, he had taken a working party up the line, where they spent several hours shoring up a section of the British defences:

> 'On our way we passed through a village that had been practically raised [*sic*] to the ground. Yet in the few houses which were inhabitable our troops were billeted. The entrances to the trenches were all labelled and it was interesting to note the names given to them, being of well known streets in London. Our work consisted of making the parapet of one of these trenches bulletproof by churning up earth from the front of the trench. It was dark when we set to and it was morning before the German machine guns were busy, traversing their front on the chance of catching a party like ours. On several occasions a burst passed straight over our heads and one felt inclined to duck. About 300 yards to our left, a farmhouse which had been shelled during the afternoon was burning away and lighting up the sky. After this work we filed out of the trench and proceeded back to our billets.'[24]

Faith remained a central and guiding influence in Don's life. The day after going up the line, he attended the 'first church parade which the battalion has had since coming out 14 weeks ago'.[25] En route to the service, held in a nearby barn, three German aircraft were spotted high above the British lines, skilfully evading all attempts to bring them down:

> 'It was a very interesting sight for as each shell bursts it leaves a little white fleck in the blue of the sky and in a moment the sky is pitted with white spots. Unfortunately "there was nothing doing" so to speak and the aeroplanes got away. All through the service our guns were booming and acted as an accompaniment to our singing.'[26]

All in all, Don considered it 'a grand little service',' a sentiment evidently shared by many of his men. 'I was censoring the platoon's letters last night,'

The Bell family, Harrogate 1912. Back row, left to right: Nancy, Billy, Don and Minnie. Front: Nellie, Smith, Dolly, Annie and Gertie. (*Bell Family Archives*)

Westminster Training College, 1910. Bell is on the back row, third from the right. (*Ph/a/3. Westminster College Photograph Album, 1893–1912. Used with Permission*)

Westminster Training College Sports Team, 1910. Bell is on the middle row, fourth from the left. (*Ph/i/1. Sporting Photographs 1910s–20s. Used with Permission*)

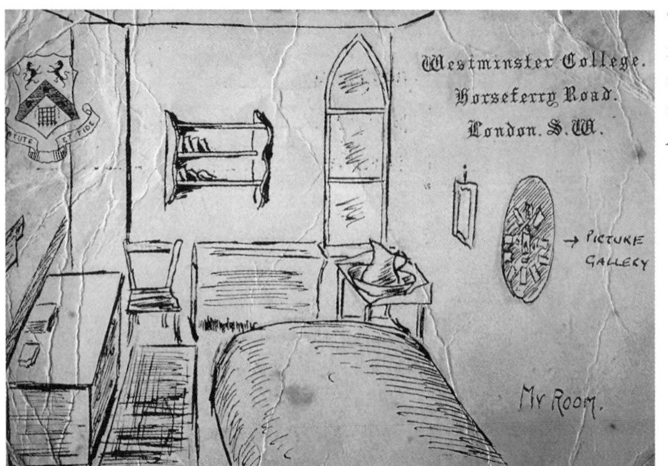

'My Room.' A sketch of Bell's room at Westminster. Sent on a postcard home soon after his arrival at the college. (*Bell Family Archives*)

Some of Bell's sports medal from Westminster College. (*Bell Family Archives*)

TOM MURRAY TRIES WITH HIS HEAD.

An exciting moment in the Bradford City v. Bradford Cup-tie at Valley Parade.. In the picture are Storer, Murray, and Bookman, of the City; Drabble, Bell, Watson, Dainty, and Garry, of Bradford.

Bradford Park Avenue versus Bradford City in the Second Round of the West Riding Senior Cup at Valley Parade on 1 October 1913. This newspaper image is the only known photo of Bell in action. (*Author*)

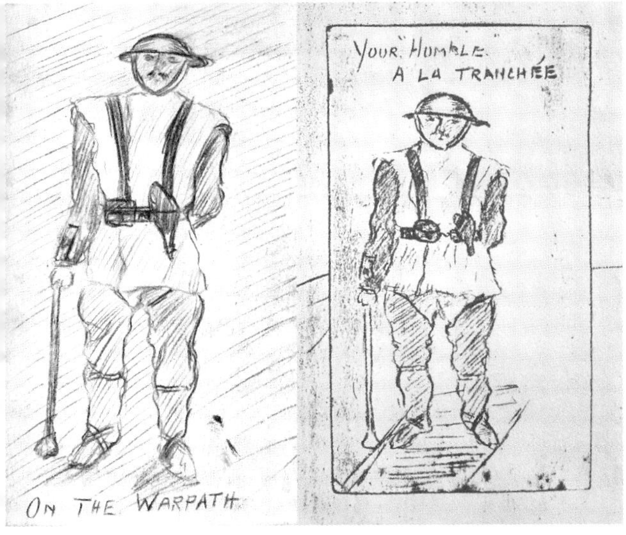

Self-portraits drawn by Bell in the trenches. (*Bell Family Archives*)

Don and Rhoda Margaret Bonson photographed before their wedding, 1916. (*Bell Family Archives*)

Painted specially for this work] [*By J. F. Campbell.*

TEMPORARY SECOND LIEUTENANT D. S. BELL DASHES ACROSS THE OPEN, UNDER VERY HEAVY FIRE, TO ATTACK A MACHINE-GUN PARTY.

During an attack a very heavy enfilade fire was opened on the attacking company by a hostile machine-gun. Temporary Second Lieutenant Donald Simpson Bell, late of the Yorkshire Regiment, immediately, and on his own initiative, crept up a communication-trench and then, followed by Corporal Colwill and Private Batey, rushed across the open, under very heavy fire, and attacked the machine-gun, shooting the gunner with his revolver, and destroying gun and personnel with bombs. This very brave act saved many lives and ensured the success of the attack. Five days later this gallant officer lost his life performing a very similar act of bravery. A posthumous award of the V.C. was made.

An artists impression of Bell's VC action at Horseshoe Trench. Published in the *War Illustrated*. (*Author*)

'In a bitter three-days fight, we oust the enemy from their underground galleries at Contalmaison.' While almost certainly staged for the camera, this stereoview slide hints at the conditions encountered by the Green Howards during the attack. (*Author*)

The ruins of Contalmaison soon after the village was captured on 10 July 1916. (*The Graphic – Author's Collection*)

POST OFFICE TELEGRAPHS

A 16.7.16

(Inland Official) (Telegrams only.) No. of Telegram 681/

Office of Origin and Service Instructions. O.H.M.S.

M.S.S. CAS.

I certify that this Telegram is sent on the service of the W.O.

Attention is called to the Regulations printed at the back hereof. Dated Stamp.

TO Bell, 87 East Parade, Harrogate Yorkshire

Deeply REGRET TO INFORM YOU THAT 2/Lt. D.S. Bell 11 Yorkshire Regt was killed in action 10th July. The Army Council express their sympathy. Please supply name address relationship next of kin

FROM SECRETARY, WAR OFFICE.

The Name and Address of the Sender, IF NOT TO BE TELEGRAPHED, should be written in the space provided at the Back of the Form.

The War Office telegram notifying the Bell family that Don had been killed in action. (*Author*)

Bell's damaged Mk I steel helmet, recovered from his grave by his brother and sent home. (*Author*)

Bell's Military Stereo 6x30 field glass. Manufactured by ***Bausch & Lomb Optical Co.*** of Rochester, New York. (*Author*)

Bell's Victoria Cross. (*Bell Family Archives*)

YESTERDAY'S INVESTITURE

Mrs. Bell carrying the V.C. won by her late husband, Lieutenant D. S. Bell. She was accompanied by her sister-in-law.

Minnie and Rhoda Bell following the investiture at Buckingham Palace on 13 December 1916. Rhoda is holding her late husband's VC in her left hand. (*Daily Mirror – Author's Collection*)

Bell's original grave at Contalmaison. Photographed in early 1917. (*Bell Family Archives*)

Contalmaison Chateau, mid-1917. Nature is already beginning to repair the scarred landscape. (*Author*)

Gordon Dump Cemetery as featured on a postcard produced in the immediate post-war period. (*Author*)

Bell's Grave today, Gordon Dump Cemetery. (*Author*)

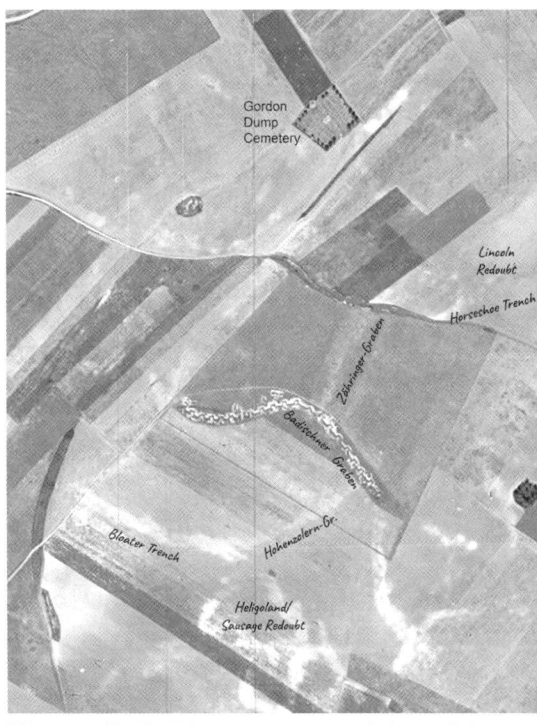

The site of Bell's VC action on an aerial photo of the 5 July battlefield taken in 1947. The remains of the *Badischner-Graben* are still clearly visible on the landscape and were not filled in until the late 1960s. (*IGN National Photo Library. 31/08/1947. Used with Permission*)

Bell's Redoubt Memorial, Contalmaison. (*Author*)

Contalmaison photographed from the south. Bell's Redoubt can be seen bottom centre. Bell was killed and buried in the green field. (*Author*)

Bell's stained-glass window, moved from Horseferry Road to Westminster Chapel, Harcourt Hill, in 1959. (*Author*)

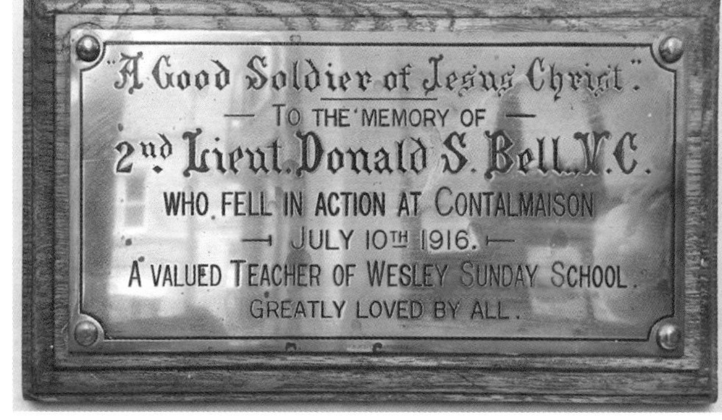

Bell's memorial plaque at Harrogate's Wesley Chapel was originally located in the Sunday School at the bottom of Cheltenham Crescent. (*Author*)

The memorial to Harrogate's three VC recipients at the base of the town's cenotaph. The paving stones either side of the plinth honour the two Great War recipients – Don Bell and his old school friend, Charles Hull. (*Author*)

he told Gertie, 'and quite a number were writing home to say what a good time they had had.' He also took part in a football match against men from 'A' Company later that day. Unfortunately, his team of officers put up feeble resistance: 'I am sorry to say we lost 4–0 and I absolutely gave away the first goal.'

On 4 December 1915, Minnie Bell, the second eldest of Don's siblings, married Thomas Wood at the Wesley Chapel. Tom, a carpenter before the war, was serving as a sergeant with the Royal Engineers and would be awarded the Military Medal in 1918. Apologizing for his absence a few days after the wedding, Don wrote: 'I am sorry I could not be there to fulfil the duties of best man.' He added wryly: 'I expect they managed without me.'

Bell spent twelve days at Fort Rompu with his battalion before marching out in heavy rain late on the afternoon of 6 December.[27] The poor state of the roads and farm tracks meant the relatively short journey to Rue Delettree, less than three miles away, took significantly longer than anticipated. By the time they completed the handover of their new billets from the 2nd East Lancs that evening, it is doubtful there was a dry foot in the battalion.

Elsewhere that day, Prime Minister Herbert Asquith received the resignation of Sir John French, commander in chief of the BEF.[28] The 63-year-old's position had become increasingly untenable, undermined by a series of costly battlefield setbacks, the political fallout from the 'shell scandal'[29] – caused by a critical shortage of artillery shells at the front – and his strained relationships with senior military and political figures. French was officially replaced on 19 December by one of his fiercest critics, Sir Douglas Haig,[30] and returned to Britain, where he was appointed commander in chief of the Home Forces, a position he held until May 1918.

The 9th Yorkshires spent the next four days in brigade reserve at Rue Delettree, providing support and a continuous stream of work parties, before moving into the line east of Bois-Grenier on 10 December to relieve the 11th West Yorkshires.[31] It was Bell's first time in the line. Rain fell incessantly for days, and the trenches were knee-deep with thick, cloying sludge that clung to his uniform and sucked at his boots.

Heavy fighting had taken place to the south during the spring and summer, most notably at Neuve Chapelle, Festubert and, in September, at Loos. However, the sector around Armentières, including Bois-Grenier, was generally regarded as a quiet one.[32] This relative inactivity, notwithstanding the occasional trench raid and persistent sniping, led to it developing a reputation as a 'nursery sector', where newly arrived battalions were sent to acclimatize them to conditions on the Western Front.

Unfortunately, the area also possessed a particularly high water table and was criss-crossed by a complex system of drainage ditches, dykes and irrigation channels. This made it ideal for farming, but terrible for static warfare.[33] Trenches flooded as quickly as they were dug. At the same time, the multitude of improvised dams and culverts thrown up to contain the water often just diverted it elsewhere along the British line or simply collapsed in heavy rain. Both sides eventually abandoned efforts to dig through the worst of the mud and instead resorted to constructing sandbag breastworks above ground. While these offered less protection than traditional earthwork trenches, they did provide some escape from the water. Nevertheless, troops in the sector waged a near-constant battle against the elements and were usually forced to endure their misfortune with nothing but sullen stoicism.

Life in the trenches was hazardous at the best of times. During the winter months it could be intolerable, as Bell would soon discover. Shortly after 8.00 am on 12 December, shellfire struck a dam close to the German front line, with fairly predictable results. A torrent of water suddenly cascaded down the slope, flooding into the British trenches. Within minutes, the disgruntled Yorkshiremen found themselves waist-deep in water and mud. Don later noted that it took two days to get their positions back to their 'former state'.[34] Trench pumps helped in such circumstances and usually succeeded in bringing water levels down. However, like most things at the front, there were never enough to go around. They also clogged frequently and required constant clearing, an irksome and filthy task in itself. Worse still, hours of backbreaking work could be undone in an instant by a single stray shell.

Bell's first spell in the trenches came to an end on the afternoon of 14 December. It had been a gruelling few days, though enemy artillery activity was relatively light and casualties comparatively few.[35] The greatest challenge had been the weather, which proved every bit as formidable as the enemy. Even so, as the battalion trudged back to billets at Rue Delettree,

Don concluded that, 'except for the difficulty of keeping dry', it had 'not been an unpleasant experience'.[36]

The next day, he wrote to his father. While letters between the pair were far less frequent than those exchanged with his mother, they often revealed a more introspective side. Don wrote candidly about the war and its hardships, unburdening himself with accounts of physical discomfort and the ever-present threat of enemy fire. He also referenced geographical locations, sometimes cryptically to evade the censors, which enabled Smith Bell to track his son's movements on the large *Daily Mail* map pinned to the wall of his study at Milton Lodge:

> 'Our company is billeted in a farm house on the main road between Bois-Grenier & a place called Fleurbaix. The trenches we occupied were due South of Bois-Grenier – about 880 yards from B.G. is the village which has been practically destroyed by shell fire. The only thing which has not been touched in the slightest is a little French chapel – not even the windows are damaged & the door is open & the candles were on the altar – yet all around it the buildings are roofless etc. The line in the district is very peculiar forming a semi-circle round Armentieres as per sketch. The result is at night you can see the Very lights going up above the enemy side.'[37]

Soon after, Bell was sent on a three-day bombing course at the battalion bomb school near Armentières. It was fairly basic stuff, covering familiar ground. Still, as he revealed to his mother, this had certain advantages: 'The best thing is I know all about the Grenadiers, having learnt them when at Darlington. So you see, this is a little rest cure for me.' He added: 'I am getting away early to visit the big town near here. "Ah meant yer" to know where I am.'[38]

No sooner had the course finished than he was back in the trenches around Bois-Grenier. The sector remained quiet, but the grim realities of war were never far away. In an unusually frank letter to his mother, written from L'Hallobeau on 22 December, Don described a macabre incident that occurred during an otherwise routine stint in the line:

> 'My platoon was stationed in a most extraordinary place for the line of trenches passed through a crowded cemetery. In the parapet were several tombstones

> & graves of English as well as French. The day before we went in a shell fell in the graveyard & burst in a vault containing five coffins, exposing them to view. The tombstone, a magnificent granite cross, was split down the middle. We had an even more gruesome experience the day before yesterday for on our way along a communication trench which was very sloppy the men unearthed five coffins, 3 of them open. The stench was fearful & we knocked off in double quick time. Although our surroundings were not cheerful it had no effect on me & I slept quite as soundly as usual.'[39]

Don was sent to L'Hallobeau for a ten-day course at the divisional school, sparing him a spell in the trenches over Christmas. Nevertheless, he was getting increasingly disgruntled at being repeatedly sent away from the battalion. 'If anyone has to be sent from C Company, as the latecomer,' he bemoaned to his mother, 'I am always dumped on it.'[40] He took particular umbrage at the demanding schedule of the course, especially when he was told he would have to work on Christmas Day:

> 'It is most awful puffer for we did sloping arms by numbers & saluting drills this morning. The hours are long too, 8.30–1, 2–4, 5–6.30. The worst of it is there is to be no break on Xmas Day or Sundays. I wouldn't have minded coming on it if we had just been going into the trenches or it had not been Xmas week, but as it is I am somewhat annoyed. However, I must make the best of it. We have very good billets here & last night I slept in a feather bed – a bit of a change that from a covered in shelf in a cemetery. Still, I should have preferred the cemetery this time.'[41]

Don did at least find some time for football while he was there. Unfortunately, the only match he participated in, against a team from the Royal Army Medical Corps, brought little cheer: 'The ground was in an awful state & the ball was like lead... We were outclassed and lost 9–0.'[42] His misery was compounded when he suffered what was almost certainly concussion during the match, the effects of which troubled him for several days: 'The fact was that last week I was off colour, my head being like lead. This was due to heading a ball that dropped a great height & not getting at the right spot.' He added: 'I am afraid I was not much use after the smack on the head but floundered along [to] the best of my ability.'[43]

Chapter 8

All Roads Lead to the Somme

Bell missed his battalion's first significant offensive operation of the war while he was at L'Hallobeau – a meticulously planned trench raid on German positions at Rue du Bois in the early hours of 1 January 1916. The operation was organized by Major H.A.S. Prior, who would go on to play a prominent role in the fighting at Contalmaison six months later, and led by Captain G.K. Thompson, who had a party 'one hundred strong' under his command.[1]

The men selected for the raid had been withdrawn to the village of Croix du Bac two weeks earlier, where they rehearsed 'in every detail on ground where trenches had been dug to represent the German works to be attacked'.[2] The divisional artillery, meanwhile, conducted several days of wire-cutting and targeted various points along the German front line, including known machine-gun positions. These efforts were scaled back two days before the assault, however, after reconnaissance reports indicated that the enemy was becoming 'more and more alert'.[3]

On the night of the operation, the raiding party filed out of their trenches at 12.15 am – faces blackened, bayonets fixed and some armed with an assortment of medieval-looking knives and clubs. After assembling just in front of their own wire, they set off towards the German positions on their hands and knees, crawling some 250 yards until they were met by a wire-cutting party, who guided them to a 5ft lane cut through the wire several hours earlier.

At 1.33 am, the British guns illuminated the night sky, bringing down a heavy barrage on the German support line and on either side of the point of entry. In an instant, the fired-up Green Howards leapt to their feet and charged down into the enemy trenches. The party on the right, led by Second Lieutenant J. Gibson, encountered minimal resistance and swept along the trench, 'killing four startled defenders with the bomb or the bayonet'.[4]

On the left, Captain Thompson's party came under attack from a small group of enemy bombers. These were swiftly driven off, with fifteen Germans later reported to have been bayoneted. A further 'five or ten more' were killed as they fled into the artillery barrage, while three prisoners were 'shot attempting to escape'. Within fourteen minutes of the raid commencing, both parties had extracted themselves from the German trenches and were on their way back to their own lines, armed with the password: 'Charlie Chaplin'.[5]

The whole operation was a resounding success, with the *Green Howards Gazette* triumphantly reporting that the battalion had 'let in the New Year in fine style'.[6] Just seven members of the raiding party were wounded – none seriously – although 'twenty-four casualties, including one officer and three other ranks killed, were sustained from hostile artillery fire' brought down in retaliation shortly afterwards. The result was all the more impressive given that the 23rd Division war diary later disclosed that 'a large number of the raiders lost direction, and only some thirty-five men actually entered the trenches'.[7]

The officers and men who took part in the operation, and a simultaneous one conducted by the 10th Northumberland Fusiliers near Ferme Grande Flamengrie, 2,000 yards further south, paraded before the corps commander on 5 January to receive his formal congratulations. Three officers were later awarded the Military Cross for their roles in the raid, while six men received the Distinguished Conduct Medal.[8] Bell rejoined the battalion at Rue Marle soon after and was eager to share details of the raid with his mother:

> 'I missed a lively time on New Year's Eve when out battalion carried out a bombing raid which was highly successful. All our men came back with several slightly wounded. The Germans retaliated by shelling our line & our company had a hot time. Two of our officers were wounded, one slightly but the other very severely. As the other officer is going on leave tomorrow we shall be shorthanded our next trip into the trenches. Fortunately we are in the support which means lying low & nothing else.'[9]

The weeks that followed were markedly quieter, with little of note occurring. On 12 January, Bell was detailed to attend yet another course, this time on sniping. 'I am beginning to get fed up with courses,' he griped. 'I seem to be dumped on them every time.'[10] Still, he had also received some good news:

> 'I expect Rhoda will have been round to tell you the good news, namely that I am coming home on leave on the 22nd of this month. I could hardly believe it when I heard but of course I made no objections. If trains & boats run to time I should reach Victoria at 4.15 on the 22nd & thus be able to get the 5.30 from King's Cross. It is very rarely that the boat is punctual & very often it sails at night. Consequently it will take me all my time to catch the midnight train, let's hope for the best.'[11]

Bell's first period of home leave since embarking for France came and went largely unrecorded, though not without incident. While standing outside the Wesley Chapel in civilian attire, he was approached by a young woman who pressed a white feather into his hand. This overzealous and – as it would later prove – bitterly ironic gesture caused considerable offence, particularly to Smith Bell. However, Don's response was not recorded.

On 27 January 1916, the Germans launched a 'terrific bombardment' on positions held by the 9th Yorkshires. One officer was wounded, and four men were killed or died of wounds during several hours of shelling, believed to have been in celebration of the Kaiser's fifty-sixth birthday. The brigade war diary notes that enemy artillery was 'unusually active'[12] again the following day, shelling the British support lines and posts in the rear. However, by that time, Bell's battalion had been relieved by the 10th West Ridings and were in billets at Rue Delettree.

On the same day, an accident occurred at the Second Army Grenade and Trench Mortar School at Téteghem, when a round exploded prematurely as it was being loaded into a Stokes mortar by a soldier from the battalion. He was later evacuated to Britain, where he spent several months in hospital before rejoining his unit in the summer of 1917. Remarkably, no other casualties were sustained in the incident.[13]

The battalion continued its rotation in and out of the line in the Bois-Grenier sector until mid-February, when 69th Brigade was withdrawn to Steenbecque for a much-needed period of rest. Over the next week or so, Bell and his comrades could look forward to regular hot food, a warm bath, clean clothing and – perhaps most importantly – a modicum of decent sleep. Pay would also be issued on their first day out of the line, which was always a welcome event, particularly when tobacco was running low. Unfortunately, the weather had taken a turn for the worse, making their brief period of respite rather less comfortable than most had anticipated. Still, as Don disclosed in a letter to Minnie, he had experienced an unexpected slice of good fortune:

> 'I believe I told you last week that we were going back to rest & here we are at last – and what a place too! We are under canvas & the camp is an absolute quagmire. We can get about in comparative comfort, however, for there are trench boards laid down. I have struck lucky again. The company commanders were to have billets in some farms nearby but two of them thought they were too far away from the mess so turned us out of our little shanty & we took over their billets… The bed is very comfortable – it is not a feathery bed but has a mattress filled with chaff – not so bad all the same.'[14]

A spell away from the sound of the guns could be remarkably restorative, but the life of an officer out of the line was rarely quiet. Even subalterns like Bell had an inordinate amount of administrative work to contend with, and most carried responsibilities that could prove both exhausting and exasperating. There were parades and kit inspections to be conducted, fatigue parties to be furnished and standing orders to be read out. Letters also had to be censored, a task one junior officer described as 'an unpleasant, impertinent duty, to be hurried over and treated as formally as possible… By constant repetition, it becomes a deadly bore.'[15] In contrast, officers were entrusted with censoring their own letters, a privilege that Don acknowledged afforded him 'more latitude' to 'reveal a little more than the Tommy' in his correspondence.

A programme of training commenced the day after the brigade arrived at Steenbecque. This included machine-gun and bombing classes, along with several route marches. Some time was also set aside for recreation, mainly in the evenings, when local *estaminets* were frequented and meagre army pay was spent on weak French alcohol and substandard food. Some soldiers occupied themselves with card games and 'Housey Housey', while serious gamblers could be found huddled around Crown and Anchor dice game boards, set up discreetly in darkened corners or back rooms.

Sport, particularly football, offered further distraction, and conditions at Steenbecque did little to dampen enthusiasm for the game. Regardless of fatigue, impromptu kickabouts would begin whenever the opportunity arose, while inter-platoon and inter-company matches always drew crowds and were fiercely contested. Bell revealed in one letter that he had been sent a new ball from Bradford Park Avenue, which no doubt increased his popularity with his men. He also sent an urgent request in another letter for his 'football togs' to be shipped out to him, complaining that 'it ruins one's clothes to play in ordinary uniform'.[16]

The battalion remained at Steenbecque until the end of the month, when it moved south to the Bruay area. Don described the journey, which took him first by rail to Camblain-Châtelain – known to the troops as 'Charlie Chaplin' – followed by a six-mile route march east to Ruitz, where both the 8th and 9th Yorkshires were billeted:[17]

> 'Reveille sounded 3.30am – a nice time to get up. As we had a considerable journey to make we did a portion of it by train. The only baggage we could have was what we could carry so you bet I had a pack full – sheepskin coat, rainproof, shirt, socks, collar, holdall, towel, blanket, along with revolver, glasses, haversack full of fruit, books etc. I had some weight. I was hugely "Red" for we did a 10 mile march at the end of it up hill & down dale…Pa will have a rough idea where we are when I say we are 8 miles S.E. of Billy's present billet. We are well behind the line & don't know what is going to happen to us.'[18]

The weather had now turned noticeably colder, with heavy snow proving of considerable hindrance to the troops. 'I am fed up with this snow,' wrote an officer with the 8th Battalion, Lieutenant Francis 'Toby' Dodgson.[19] 'It does make everything such a mess and it came down like anything just as we were getting here.'[20]

As the British expanded their operational responsibilities along the front, brigade headquarters made preparations to take over a section of the line held by French troops east of Carency. In the early hours of 6 March, Bell and his battalion, accompanied by two sections of the Machine Gun Company, trudged through a snowstorm to relieve elements of the 17th French Division in the Souchez sector. The trenches were in a deplorable state, offering little protection from the elements or enemy artillery fire. Bell later admitted that this spell in the line was 'the roughest passage' he had endured since arriving in France, adding: 'If you knew the time we have been having this last two days you would wonder how we are alive to tell the tale.'[21] He went on:

> 'The trenches we occupied were just an open ditch on a moor & were everywhere knee deep in chalky sludge. I was first man in & consequently found all the holes & in one place went up over the thighs. There were no dug outs in the line at all – in fact in places men were exposed from the waist. There were 2 or 3 holes in the rock in a communication trench behind & there we cowered all day while the artillery let drive with aerial torpedoes, sausages [large-calibre high-explosive German ordnance] & shells. The first day we had a tin of salmon, a loaf of bread & a bottle of cold tea for two officers. No fires were allowed, no blankets. So you can imagine what a time we had. This part of the line has witnessed some [of] the bloodiest fights of the war. The villages just behind have not one stone or another. There is not 5 sq yds of space which has not a shell hole in it. The result is the ground is so churned up as to form a spongy mass.'[22]

After two arduous days in these positions, the 9th Yorkshires were relieved at 11.00 pm by the 10th West Ridings and withdrew to billets at Villers-au-Bois. Conditions there were scarcely better:

> 'The billets behind the line are awful & are cellars of ruined houses. There are no shops & it is impossible to get any extras. Now is the time when parcels really will be most acceptable [and] my only hope is that the division do not stay long here. The other region we were in was an absolute picnic compared to this & it strikes one that the former place is a kind of breaking in section for divisions newly sent out... The weather here has been very bad, to make matters worse, & one night it snowed incessantly [but] fortunately the Bosche [*sic*] opposite were very friendly or we should have had it hot. I shall have some tales to tell next time I come home which I hope will be before long.'[23]

The 8th Yorkshires also moved into the line during the early hours of 8 March, slogging their way up from Gouy-Servin to occupy positions next to their sister battalion. With the 11th West Yorkshires on their right, the brigade now held a front extending from the River Souchez in the north to a position known as 'Ersatz Trench' in the south.[24] All units faced similar hardships. 'It was the most arduous time we've had and a great change from anything we'd done before,'[25] wrote Lieutenant Dodgson. 'The trench was the worst thing I've seen... You couldn't cook, except on a Tommy cooker and water was very scarce, as every drop had to be carried from the bottom of that damned hill.'

On 10 March, Brigadier General F.S. Derham relinquished command of 69th Brigade and returned to England.[26] He was replaced by Brigadier General T.S. Lambert, the popular commanding officer of the 2nd East Lancs. Lambert would lead the brigade through the bloody fighting on the Somme and at Third Ypres, before taking it to Italy in late 1917. In May 1918, he returned to France to take command of the 32nd Division and was succeeded by Lieutenant Colonel A.A. Beauman.[27]

On the day Derham left the brigade, Bell and his battalion moved into the support trenches at Notre Dame de Lorette.[28] Situated on the summit of a prominent ridge, the position dominated the Douai Plain and the town of Arras to the south and had been heavily fought over during the early phase of the war. It was eventually secured by the French during the Second Battle of Artois in May 1915, but it took until September and the Third Battle of Artois to complete the descent into the valley below.[29]

Bell had recently received news that Private Charles Hull, a Harrogate man serving with the 21st Lancers on India's North-West Frontier, had been awarded the Victoria Cross for rescuing a wounded officer during skirmishing with tribesmen at Hafiz Kor on 5 September 1915.[30] The 25-year-old was a former classmate, and the two had known each other well. 'I was delighted to see that Charley Hull had won the V.C.,' Don wrote to his mother. 'He was in the same class as I was at St. Peters and a chum of mine.' He added, almost prophetically: 'I suppose Pa won't be satisfied till I have got the V.C., D.S.O. and any other things possible. Like Asquith, we must Wait and See.'[31]

Smith Bell had more pressing concerns. On 14 March 1916, he found himself in court representing the Wesley Chapel, which stood accused of breaching wartime lighting restrictions. Evidence was heard that light had been seen shining through the chapel's windows onto the street outside. As an official, Smith Bell was summoned to appear before the Harrogate Bench, where he argued that the interior lights were shaded and could not have been visible, as the windows were 'opaque'. The inspector, Mr Jackson, contended that the windows were not opaque enough to block the light and, although blinds had been fitted, they were not drawn. The bench agreed with the inspector and imposed a fine of ten shillings on the chapel.[32]

By mid-March, Bell was back at Camblain-Châtelain, where the battalion spent ten days engaged in company and platoon training.[33] He penned several letters during this period, including a notably lengthy one to his mother in which he spoke candidly about some of the more harrowing aspects of life in France:

> 'We are well back from the firing line now as we were marching practically all day yesterday. You will be pleased to hear that we do not go back to the same sector as we were relieved by another division when we came out. I don't suppose we shall go very far from it and anywhere in this neighbourhood as the fighting is fairly hot[,] still we can assure ourselves that conditions can be no worse. I am glad for somethings that we went to that part of the line, for it showed both us and the men that we had had a picnic where we were before and now anything other will be quite pleasant, comparatively speaking. It has been a lively quarter and no mistake for there are bodies or what is left of them lying in all directions. The last two days we were there it was very hot and the place "miffed" horribly, consequently I am glad to be away.'[34]

A short time later, the 9th Yorkshires moved into the line in the Angres sector.[35] The trenches in this area represented a marked improvement over those previously occupied, featuring well-constructed communication trenches and numerous deep dugouts cut down into the chalk. It was also relatively quiet, allowing officers and men time to read and write letters during breaks in their trench routine:

> 'We are still in the trenches but have not much more to do as things are pretty quiet with the exception of rifle grenades. As only about 1 in 20 land into the trench you have a pretty good chance. The weather has been the worst as it has been very cold and wet. This afternoon though it is gloriously fine and it is quite pleasant to sit out in the sun.'[36]

The battalion had now been issued with the new Mk I steel helmet. Designed to offer greater protection against shrapnel and other battlefield hazards, the 'Brodie' helmet was a significant improvement over the soft cloth service caps previously worn.[37] However, while they did dramatically reduce the number and severity of head injuries, they were uncomfortable to wear and offered limited defence against direct bullet impacts and larger shell fragments. Nevertheless, the helmet's practical advantages were quickly recognized, and it soon became an indispensable item of equipment. After receiving his helmet, Don wrote:

> 'You would laugh if you saw me in my trench rig out for added to my former kinds of apparel is a shrapnel helmet which is not close fitting like the French but more like an inverted soup plate. I can assure you I look a regular desperado when I turn out. The one disadvantage about them is their weight for after a time your head begins to ache. Still one can endure that as long as the head is protected.'[38]

A rare moment of drama occurred on the afternoon of 2 April, when a French aeroplane was shot down over British lines by German anti-aircraft fire. After watching it crash 600 yards to the rear of the 8th Yorkshires' headquarters, men from the battalion rushed to the wreckage in search of survivors. Sadly, by the time they got there, they found the pilot already dead and his observer mortally wounded.[39]

During another relatively quiet moment in the trenches, Don took the opportunity to raise 'the momentous question of football on Sundays' with his parents. As devout Methodists, they discouraged all secular activities on the Sabbath, including organized sport. But the upheaval of war was challenging long-standing social and religious conventions, and Don sensed it was time to tackle the issue directly. 'I thought you would be averse to the idea, but I do not like to do things on the sly,' he told them. 'I felt I was justified in playing out here where conditions are absolutely different from what they are in England.'[40] He continued:

> 'A platoon is like a family and as Sunday is the only day free from parade we have to find something for the men to do. They are not allowed to leave the billets so football is about the only thing which provides healthy amusement and recreation for the men. Assuredly it is much better to do this than be in the huts playing cards and gambling. We of course have to enter into it with the men. You might say in reply why not give a Bible Class, well in reply I might state that there is church parade on the morning and optional service at night. These games of football are only indulged in when the battalion is in rest billets which does not occur more than once a month. I have erred in good company for young Erob is a local preacher and he has played in the

> same games as I have done. We told the chaplain last Sunday and he seemed to take it as the natural state of things.'[41]

It was one of the many quirks of the war that Don and Billy Bell spent most of their time in France within earshot of one another. Indeed, the brothers were reunited on several occasions:

> 'You say Billy wrote complaining that I have never answered his letters. I can't very well when I never get them. He came to see me last Sunday so we thrashed the matter out & it appears that he places 11th West on instead of 9th so what can you expect. He didn't even know the cap badge & was on the look out for men wearing the white horse badge. I had a shock when I saw him sidling across a field towards me… It appears they had only been 3 miles away instead of 7 as I had thought & every day his waggon had passed within ½ a mile of our former billet. After a great deal of persuasion I got him to stay to tea – (he wanted to engage a private room somewhere – an impossibility in France) & then I sent him on his way home.'[42]

On 9 May 1916, Don and Billy's cousin James Simpson, a second lieutenant with the 173rd Tunnelling Company, Royal Engineers, was killed at the front.[43] The recently engaged 28-year-old, who had initially served with the 1/5th West Yorkshires before receiving a temporary commission in September 1915, was a well-known local rugby player and one of the founders of the Harrogate Old Boys Club. Don, understandably saddened by news of his cousin's death, wrote to Dolly shortly afterwards:

> 'I was very sorry to hear about Jimmy Simpson's death. There's no doubt about it Harrogate is losing some good men. The worst of it out here [is] you never know when you are going to get one. A man may come through a hail of bullets & remain untouched & then be caught by a stray one the behind the lines. However one takes things as they come & never worry.'[44]

By coincidence, Jimmy's younger brother, George Simpson, was serving nearby with No. 3 Brigade Company, Australian ASC, and was granted time off to attend his brother's funeral at Nœux-les-Mines.[45] George was in Australia when war broke out and enlisted almost immediately. After a brief spell in Egypt, he sailed for France, arriving just four weeks before Jimmy's death. Billy Bell, who was also billeted nearby, was carrying some men from his company up to the front one night when he heard of the tragedy that had befallen his cousin. He visited Jimmy's grave shortly afterwards:

> 'I walked to see Jimmy's grave last night & found rather a nice cross erected, painted black & white which is the pattern for officers. Have not troubled to go to the company when I knew George had attended the funeral. I had all information first hand from the sergeant.'[46]

Billy had a narrow escape himself that night. 'Unknowingly, my mate and I were strolling about "suicide corner" near Loos,' he wrote, 'when four bullets came in quick succession, one passing between our heads. Needless to say we scattered and felt very thankful we had suffered no hurt.'[47]

George Simpson was not so fortunate. After receiving a commission and being posted to the 6th Battalion, AIF, he was badly wounded at Glencorse Wood in September 1917 and evacuated to a hospital in England. Although he eventually recovered and rejoined his unit, he was killed in action near the village of Foucaucourt on 23 August 1918.[48] He was 26 years old and is buried at Cerisy-Gailly Military Cemetery, six miles south-west of Albert.

In May 1916, Parliament passed the second Military Service Act, extending conscription to married men aged between 18 and 41.[49] Stricter conditions were also imposed on those seeking exemption, significantly narrowing the grounds for appeal before local Military Service Tribunals. Public opposition to conscription – first introduced in January – remained widespread, fuelling industrial unrest and exacerbating political tensions across the country. Reactions at the front were mixed. In the face of mounting casualties and the failure of the 'Derby Scheme' urging men to voluntarily attest for future military service,[50] some soldiers accepted it as a necessary measure, although the first conscripts would not reach France for several months. Others, particularly early volunteers and regulars, viewed conscripts with suspicion, questioning both their commitment and their ability to endure the rigours of trench warfare. Don's frustrations with events at home are evident in a letter to Minnie, written in the aftermath of a Zeppelin scare over Yorkshire:

> 'You seem to have had a lively time with the Zepps but really out here we never bother much about them. If it were not for the thought of the women

> & children getting damaged I should be rather pleased that they do come for they seem to be the only things which [make] some of the so-called men of England realise that there is a war at all. I get absolutely fed up when I read of the tribunals & strikes when there are lads who have been in the army 20 months & are only just 19 now after 9 months trench life.'[51]

It was a sentiment likely shared by many of his comrades, one that is perhaps understandable given the cumulative strain endured by men at the front. Don rarely let his veil of resilience slip, but a letter home at the end of April offers a rare glimpse into his state of mind. He had been feeling 'off colour' for several days and was now in hospital with influenza. 'By Jove, it must be grand in Harrogate just now,' he told his mother. 'I have been feeling quite homesick recently. I have either been so worn out or down in the dumps.'[52]

After being discharged from hospital, he was sent on a bombing course and missed a turn in the trenches. By the end of the month, he was back with his battalion and finally had some positive news to relay:

> 'Just a line to let you know that leave is going again & that I am definitely coming on the 2nd. Rhoda will send you all particulars about the ceremony so will say nothing about that. I intend paying Billy a visit before I come so will be able to tell you how he is faring. I received a herald today – the first for a month. Will Pa see Mr Irving at once to see if he will be able to officiate on the Monday. I'll let Rhoda know. I shall not be able to secure an answer back here in time so must leave it to you.'[53]

Don did stop off to see his brother en route to Boulogne, and the pair spent several hours together before parting ways. It came just days after Britain's Grand Fleet had clashed with the Imperial German Navy's High Seas Fleet at Jutland, though details of the engagement were still emerging.[54] Sadly, it would prove to be the last time the brothers would ever see each other. 'I wish I was coming, too,' Billy lamented, 'for Harrogate must be looking well just now, but I must wait two or three months yet before my turn comes.'[55]

On Monday, 5 June 1916, Donald Simpson Bell married Rhoda Margaret Bonson at the Wesleyan Chapel in Kirkby Stephen. News of the wedding appeared in the *Harrogate Herald* two days later:

> 'At the Wesleyan Church, Kirkby Stephen, on the 5th inst., by the Rev J R Irving, of Harrogate, and the Rev A Smith, of Kirkby Stephen, Donald S Bell, Second Lieutenant 9th Yorkshire Regiment, British Expeditionary Force, younger son of Mr and Mrs Smith Bell, Harrogate, to Rhoda Margaret Bonson, elder daughter of Mr and Mrs Bonson, Kirkby Stephen, Westmorland.'[56]

Only one photograph of the couple is known to survive, taken some time before their marriage. It is a remarkable image, conveying a sense of exuberance rarely seen in photographs from the period. Don, immaculate in his uniform, Sam Browne belt and perfectly polished boots, holds Rhoda protectively with one arm, her elbow resting gently on his thigh. Both beam with happiness. The photograph captures a fleeting moment of tenderness and affection, far removed from the death and squalor of the trenches.

There was other significant news that day. Just hours after the wedding, the Royal Navy armoured cruiser HMS *Hampshire* struck a German mine in heavy seas off the coast of the Orkney Islands, going down with the loss of more than 730 lives. Among the dead was the Secretary of State for War, Lord Kitchener, who had been on his way to the Russian port of Archangel for a secret meeting with Tsar Nicholas II. News of Kitchener's death shocked the country and dealt a serious blow to morale. 'At that moment we felt like the war was lost,' wrote one Northumberland Fusilier. 'The man whose rallying call we had all heeded was gone. It felt like a blow far greater than any German victory on the battlefield.'[57] Another likened it to 'losing his favourite uncle when he was a child'.[58]

Bell's time at home was, sadly, all too brief. After a short honeymoon at the Croft Hotel near Darlington, he bid farewell to his wife and family and was soon on a troopship cutting back across the Channel to France and the war. Unbeknown to them all, it was a journey from which he would not return.

The 9th Yorkshires saw out the next couple of weeks in the Souchez sector. At dawn on 24 June 1916, they formed up and marched eight miles east to Berguette, where they entrained for Longueau, a major rail junction on the outskirts of Amiens. The gruelling journey south, which took more than seven hours, was followed by a nine-mile route march to billets at St Sauveur, where they arrived after dark. Here they would remain for the next six days. Any man with lingering doubts as to the purpose of their move south 'could scarcely have now failed to realise that business was intended after viewing the scenes of activity which prevailed at Longueau sidings'.[59]

In the opinion of Lieutenant Colonel Holmes, the 9th Yorkshires were now 'a very fine fighting unit indeed, and perhaps more efficient and better disciplined than at any other time'.[60] Ten months in France had hardened them to the discomforts of trench life, and most had developed a cautious indifference to shellfire. Yet Holmes also noted that the battalion was

> 'to all intents and purposes the same personnel who had assembled at Frensham Camp nearly two years before... The casualties which it had incurred during this period were negligible and had been replaced by drafts. Officers, NCOs and men were all better acquainted with one another. Sections and platoons all possessed capable and experienced leaders.'[61]

On the cusp of the British offensive on the Somme, Bell wrote home from St Sauveur. 'You will have seen from the papers that things are livening up on this front,' Don told his mother. 'The Russians & the Italians are getting into their stride & it won't be so long before we join in the show.'[62] How right he would be.

Chapter 9

The Mouth of Hell

'Behold, the day of the Lord comes, cruel, with wrath and fierce anger, to make the land a desolation and to destroy its sinners from it.' – Isaiah 13:9

On 24 June 1916, a balmy Midsummer's Day, British batteries along a 16-mile front roared into life, unleashing a bombardment of such unprecedented ferocity on the German positions that few envisioned anything but their complete destruction. Field guns positioned a thousand yards behind the line, and 'heavies' further back still, pounded the enemy defences relentlessly, sending over more than 150,000 shells every twenty-four hours.[1] The bombardment was so intense that recoil buffers on some guns failed under the strain.[2] Unfortunately, the weather then began to deteriorate. Rain fell incessantly, and mist and low cloud hung over the battlefield, grounding observation balloons and the aircraft of the Royal Flying Corps. Eventually, the decision was taken to push the attack back forty-eight hours to 1 July, and the deadly overture continued in all its fury.[3]

Despite the prodigious number of shells expended, the onslaught failed to deal adequately with the German front line.[4] Their second line, which lay beyond the effective range of most British guns, was barely affected at all. Most enemy soldiers sat out the bombardment in deep dugouts carved down through the chalk, some more than 30ft below ground and impervious to all but the heaviest of shells.[5] Here they endured a fearful ordeal, but with ample time to prepare, they emerged largely unscathed when the barrage lifted at 7.30 am on 1 July – unleashing seven days of pent-up fury on the unsuspecting British infantry advancing towards them. High-explosive (HE) shells were essential for targeting such defences; however, they were in chronically short supply. In fact, only one-third of all shells fired before the assault were HE rounds. The remainder were predominantly shrapnel – deadly against infantry in the open, but useless against men sheltering underground.[6] Fuses were also unreliable, frequently failing to ignite or even falling off in flight. As a result, large swathes of barbed wire remained uncut, and a sufficient number of German artillery positions survived counter-battery efforts to inflict brutal losses once the attack commenced.[7]

It was not until the opening day of what would soon become known as the Battle of the Somme that Donald Bell and the 9th Yorkshires moved out of St Sauveur and began their journey towards the front. They had been there six days, waiting anxiously for news that the great offensive had finally begun. Don, his hair freshly cropped, wrote to his mother three days before the battalion moved out:

> 'Since writing last we have had a move of considerable distance & I am now miles away from Billy. This is a fine part of the country & as we are 20 miles from the line there is no need to worry about how I am faring... I'm afraid I have no news this time as there is absolutely nothing fresh. I am feeling very fit – I had a hard game of football the other night & was in great form – felt as fresh as paint at the end too... You would laugh if you could see me now for I look as if I had first been released from Armley Jail. I told you I intended having my hair cropped short & I've done it – had the shears all over. You can keep cool & it is much better for wearing the steel helmet.'[8]

The battalion's first stop was the village of Baisieux, where they bivouacked for the night in woods to the west.[9] Here, disquieting stories began to circulate that things had not gone to plan. 'Many rumours began to be received of the doings of the 8th Division,' writes the regimental history, 'the wounded from which were being evacuated along this route.'[10]

Particular interest was shown in the fate of 70th Brigade and its three battalions of Yorkshiremen, for they had once formed part of 23rd Division. Advancing up the southern slope of Nab Valley, north of Ovillers, on 1 July, the leading waves had succeeded in getting across no-man's land and into the German trenches. Some even got as far as their third line. By mid-afternoon, however, all gains had been lost, and the survivors forced to withdraw back over a no-man's land now strewn with rows of khaki-clad dead.[11] 'Not only were they laying side by side,' recalled one German machine-gunner, 'but also one on top of the other.'[12] The brigade's losses were the heaviest in the division, with seventy-three officers and nearly 2,000 men reported killed, wounded or missing. By the end of the day, the 8th Division had sustained more than 5,000 casualties, including close to 2,000 dead. 'The experience of this day had been bitter,' lamented the divisional history, 'and its losses terrible.'[13]

The 9th Yorkshires left Baisieux late in the afternoon of 2 July and marched east towards Albert. Though little more than six miles away, their progress was severely hindered by the heavy volume of traffic on the approach roads and tracks. Prisoners were among those streaming west, but their number was far exceeded by the flow of ambulances and the long lines of walking wounded. By the time the battalion finally reached its bivouac site, three-quarters of a mile north-west of Albert, shortly before midnight, a temporary lull had settled over much of the battlefield, broken only by sporadic bursts of artillery and the muffled rattle of machine-gun fire.[14] Even so, as Bell and his men tried to snatch a few hours of fitful sleep, most knew the respite would be short-lived.

The order to move, which came at 9.30 am the next morning, brought with it the realization that 'at last the supreme test was at hand':[15]

> 'There was a thrill of expectation and excitement in the air. All were outwardly calm and impassive, but yet each man felt within him that the days were at last approaching for which all had been preparing for over two years, and each asked himself whether he would be able to apply all the lessons which he had learnt and fulfil all his highest hopes and ambitions.'[16]

The 8th and 9th Yorkshires now moved forward to occupy a 'position of readiness' along the Tara–Usna line, some 1,500 yards west of La Boisselle.[17] There, they were issued hot meals and awaited further orders. Extra rations and ammunition had already been drawn, and rifles and kit inspected. Most of the men had also written letters home, particularly those wishing to put their affairs in order before going into battle. To the south, at Bécourt Wood, the soldiers of the 10th West Ridings and 11th West Yorkshires were doing likewise.

As they marched through Albert towards the front, Bell's battalion had passed the town's once-majestic cathedral and its fabled Golden Virgin. Leaning precariously from the shell-damaged dome of the great basilica, the gilded Madonna 'seemed as though she might at any moment fall with the Holy Infant in her arms.'[18] Legend had it, wrote one soldier apocryphally, that 'when the Virgin falls, the war will end'.[19] Observation balloons and aeroplanes, 'busily engaged upon their morning tasks', dotted the sky, while British batteries 'revealed themselves at every turn'. It was 'such a concentration of artillery', wrote historian Colonel H.C. Wylly, 'as the world had never before witnessed, nor indeed had deemed possible'.[20]

The task of capturing La Boisselle had been assigned to Major General Ingouville-Williams's 34th Division, composed predominantly of locally raised 'pals' battalions like the Tyneside Irish and Tyneside Scottish, the 'Grimsby Chums' and the 16th Royal Scots – 'McCrae's Battalion'. Their assault was supported by the detonation of two huge mines – Y Sap and Lochnagar – dug beneath German strongpoints on either side of the village.[21] Unfortunately, the enemy knew they were coming, alerted to the attack when an underground Moritz listening device intercepted a British message sent in the early hours of 30 June.[22] Within ten minutes of leaving their trenches, 80 per cent of the men in the lead battalions had been scythed down by German machine-gun fire. By the end of the day, the 34th Division had lost 6,380 men – the heaviest casualties sustained by any division along the entire sixteen-mile front.[23]

Despite these monstrous losses, some gains had been made south of the village, where a foothold almost 1,000 yards deep had been secured. Isolated parties had even reached the village of Contalmaison, but were either captured, killed or forced to withdraw towards Scots Redoubt, where mixed groups now clung to positions alongside elements of the 21st Division.[24] A smaller force, no more than 150 men in total, was also holding out between the still-smouldering mine crater close to the German strongpoint at *Schwabenhöhe* and the southern edge of the village.[25] It was imperative that these gains, modest though they were, were consolidated, and fresh troops sent to relieve the exhausted men who had so valiantly secured them.

At 5.30 pm on 3 July, 69th Brigade was ordered forward to take over positions from 101st Brigade and 102nd (Tyneside Scottish) Brigade. Soon after, the 9th Yorkshires and the 11th West Yorkshires left their positions on the Tara–Usna line and set off towards the beleaguered troops entrenched on the far side of Sausage Valley, the name given by the British to one of two shallow valleys near La Boisselle, the other being Mash Valley.[26] As they crested the hill, Bell caught his first sight of the battlefield spread out before him. Clouds of dust and smoke shrouded much of the village as the British guns continued to pound the survivors of the German garrison still holed up in the rubble, while the detritus of previous assaults lay scattered across the scarred landscape. Avoca Valley was littered with the fallen, men of the Tyneside Irish Brigade, cut down in long lines as they advanced down the exposed slope, still hundreds of yards behind their own front line.[27] To their left, the kilted dead of the Tyneside Scottish, played into battle by their pipers, lay strewn across the ground. Partially obscured by the smoke drifting over Sausage Valley, the bodies of many more men could be seen, so numerous they formed a grotesque path tracing the direction of the British advance less than forty-eight hours earlier.[28]

After labouring their way up the steady slope, rain now lashing their faces, the Green Howards finally linked up with the remnants of the 101st Brigade and 102nd (Tyneside Scottish) Brigade that evening.[29] The area had been pulverized by the artillery of both sides, and corpses were scattered everywhere in various states of mutilation, bloated and putrescent. The ghastly scene, matched only by the fetid stench, turned the stomachs of the newly arrived troops. But few of those who had lived cheek by jowl with the dead for days noticed anymore. Most of these sun-blackened bodies would remain where they had fallen until Contalmaison was finally secured on 10 July. The horrific scene was described by Lieutenant Colonel H.R. Sandilands in the divisional history:

> 'The relief in the front trenches, which were taken over after dark, was a trying experience for troops unaccustomed as yet to the aftermath of a great battle. The trenches, now battered to bits, had been the scene of terrific carnage. They were literally choked with dead bodies, which it was impossible to avoid treading underfoot, and the sensations of the young soldier, conscious that he was shortly to enter battle himself, can be readily realised.'[30]

By midnight, the 9th Yorkshires had completed the relief and were now deposited along a battered length of trench known to the Germans as the *Badischer-Graben*. On their right, the 11th West Yorkshires held positions extending to the northern edge of Round Wood, including Scots Redoubt. Further back, the 8th Yorkshires and the 10th West Ridings were in reserve at Tara Hill and Bécourt Wood respectively, though both units would soon enter the fray.[31] Brigade headquarters was also at the Tara–Usna Line, where a wire had been laid to connect it with Scots Redoubt. Unfortunately, this was frequently cut by shellfire, and runners became 'practically the only means of communication',[32] despite it taking ninety minutes to dodge through the shells and choked trenches to reach the forward positions.

Bell and his battalion were now in close proximity to the enemy, and shortly after taking up position, bombing parties were sent out to engage them. A night of chaotic, close-quarters fighting ensued, as both sides fought tooth and nail in the dark to gain the upper hand. By 9.30 am, the determined Yorkshiremen had managed to systematically edge their way up a German communication trench, the *Zähringer-Graben,* towards Point 49 and Lincoln Redoubt beyond. The 11th West Yorkshires, meanwhile, were also making headway and had almost reached Point 56 by 2.00 pm, when they were driven back by strong German counter-attacks.[33]

Localized fighting and considerable bombing activity continued throughout the remainder of the afternoon.[34] While sporadic exchanges of rifle and machine-gun fire persisted after dark, punctuated by the occasional boom of exploding Mills bombs and *Stielhandgranaten* in the contested trenches, there was little movement of the line. Instead, the two battalions of Yorkshiremen paused and took breath. As they did, three companies of the 8th Yorkshires spent the night digging a new communication trench from the old British front line to Sausage Redoubt, with the remaining company placed at the disposal of the officer commanding their sister battalion.[35]

There had been a noticeable deterioration in the weather over the preceding twenty-four hours, and 'frequent and heavy showers' continued to blight the men in the line. This culminated in a thunderstorm on 4 July that lasted all afternoon,[36] making the ground underfoot 'hourly more unfavourable to infantry attack'.[37] Heavy cloud cover also rendered air support particularly difficult, though the Royal Flying Corps did succeed in getting some machines up to register targets for the artillery.[38]

By the early hours of 5 July, 69th Brigade's front measured approximately 1,000 yards in length. On the left, the 9th Yorkshires held the *Badischer-*

Graben from a position close to the junction with the *Steinmann-Weg*, while the 11th West Yorkshires held positions stretching to the junction of Round Wood Alley and Dingle Trench on the right. Bombing parties from the 10th West Ridings, recently arrived from Bécourt Wood, were now engaged in the southern portion of Horseshoe Trench, attempting to advance toward Point 74. Meanwhile, parties from the 11th West Yorkshires were pushing north towards Point 35 from Scots Redoubt.[39]

Progress was slow, with the enemy fighting with 'skill and determination'[40] for every section of smashed and debris-choked trench. The skirmishing culminated in a fresh assault by the two battalions just before dawn, resulting in gains on the right and centre of the brigade front. Unfortunately, on the left, the 9th Yorkshires encountered stiffer opposition and were unable to advance any further up the *Zähringer-Graben*. Nonetheless, by first light, it was reported that Points 35 and 56 had been secured and were being consolidated in anticipation of likely German counter-attacks.[41]

The first of these attacks came three hours later, targeting the 10th West Ridings, who had fought their way up Horseshoe Trench to within 50 yards of Point 74. Despite offering determined resistance, they were forced to relinquish much of the ground they had taken, but eventually managed to regroup after a temporary trench block was erected.[42] Soon after, the 11th West Yorkshires were also subjected to strong thrusts from the north and north-east, and were driven back towards Scots Redoubt, where they finally repulsed the enemy with assistance from a company of the West Ridings that had been held there in reserve.[43]

With their toehold in Horseshoe Trench all but eradicated, brigade headquarters set about planning a renewed assault with the intention of capturing it once and for all. 'Hitherto, trench-bombing tactics had been employed,' records the divisional history, 'and an attack across the open by the 10th Duke of Wellington's, 8th Yorkshire Regiment, and 9th Yorkshire Regiment in line from right to left was now to be attempted.'[44] At 2.45 pm, orders were dispatched to the two battalions of Green Howards and the 10th West Ridings, instructing them to prepare for an assault scheduled for 6.00 pm that evening.[45]

The objective of the attack was the capture and consolidation of several points along Horseshoe Trench and positions extending eastward to the junction with Birch Tree Alley, south-west of Peake Wood.[46] An hour-long artillery bombardment was to precede the assault. Each battalion was to be allocated four machine guns and two Stokes mortars, with another four machine guns and two Stokes mortars held in reserve at Scots Redoubt, ready to be brought into action on request. Throughout the attack, the 9th Yorkshires were to maintain contact with 56th Brigade, 19th (Western) Division, on their left, while the 10th West Ridings were to keep in touch with elements of 52nd Brigade, 17th (Northern) Division, on their right. All reports were to be sent to advanced brigade headquarters at Scots Redoubt.[47]

At around 3.00pm, 9th Battalion headquarters moved forward from Chapes Spur to Heligoland, a distance of around half a mile, where it was re-established in an old German dugout.[48] It now sat less than 300 yards behind the battalion's positions, enabling quicker communication, particularly for runners. It also brought it closer to advanced brigade headquarters, which was equipped with a wireless set capable of sending and receiving coded messages. The 8th Yorkshires had also been ordered forward to relieve the 11th West Yorkshires, establishing their headquarters at Scots Redoubt.[49]

The preliminary bombardment began at 5.00 pm and was deafening in its intensity, causing the ground to heave and groan under the feet of the waiting Green Howards. Artillery support had initially been assigned to the 23rd Divisional Artillery; however, they were re-tasked to operations west of Fricourt, and responsibility passed to the guns of the 34th Divisional Artillery.[50] Designed to soften up the enemy before the assault, the hour-long bombardment would increase in intensity fifteen minutes before zero hour. It would then lift onto targets behind the main objective in an effort to disrupt enemy reinforcements.[51]

Despite heavy shellfire and driving rain, British efforts to dislodge the enemy continued throughout the afternoon. These culminated in coordinated advances along the southern section of Horseshoe Trench and across open ground, forcing the Germans back towards Point 74. Realizing their line of retreat had been cut off, a large number of the enemy now chose to surrender.[52] By 5.45 pm, British troops had fought their way doggedly up some 500 yards of Horseshoe Trench from Round Wood Alley, retaking Point 56 and capturing more than eighty prisoners at the point of the bayonet.[53] The Germans were badly shaken and could be seen fleeing east in considerable numbers. With a significant portion of Horseshoe Trench now under their control, brigade headquarters hastily recast its plan of attack:

> 'Orders had been issued for an attack over the open to be delivered by the 9th and 8th Green Howards and the 10th Duke of Wellington's, but owing to the progress made on the right these orders were cancelled at the last moment as regards the two last-named battalions, and the attack was to be made at 6.pm by the 9th Battalion The Green Howards only.'[54]

These revised plans were soon thrown into serious jeopardy when a concealed German machine-gun position was identified on the battalion's left flank. With zero hour fast approaching, a party of men led by bombing officer Lieutenant John Gibson was sent to destroy the gun.[55] Unfortunately, their efforts quickly unravelled when they were spotted by the enemy, who immediately opened fire. Twenty-three-year-old Gibson, a native of Kingston-upon-Hull, was hit and tumbled to the ground, 'riddled with bullets'.[56] Leaderless and unable to advance any further, the remaining members of the party were forced to scramble back to safety and abort their mission.

With zero hour now imminent, there was little more to be done but hope that the final, brutal salvos of the British bombardment would silence the gun before the Green Howards emerged from their trenches to begin their attack.

Chapter 10

These Stirring Times

A thick air of trepidation hung over the British trenches before the attack. Given the ferocity of the bombardment now pounding the German positions, some of the Green Howards risked quick peeks across no-man's land to assess its effects. However, Horseshoe Trench was barely visible through the dense cloud of cordite smoke and dust drifting across the battlefield. It certainly appeared mightily impressive as the guns boomed towards their final furious crescendo. The noise was 'deafening', recalled one soldier, 'like an unstoppable express train hurtling through the night'.[1]

At precisely 6.00 pm, a momentary pause descended over the battlefield as the British gunners adjusted their sights onto the next line of enemy defences. Seconds later, the tension was shattered by the shrill blasts of trench whistles breaking out along the line, and the men of the 9th Yorkshires – urged forward by fired-up platoon and section commanders – scrambled out of their trenches and launched themselves into the maelstrom.

Their first objective, Point 49, lay just over 200 yards ahead. Once taken, they were to advance east along Horseshoe Trench to secure Points 79 and 00, before linking up with the remainder of the brigade at Point 38.[2] If the attack was successful, the entire length of Horseshoe Trench would be in British hands by nightfall, providing a vital springboard for an assault on the strategically important village of Contalmaison, 1,200 yards to the east. It was, without question, a tough task, but the earlier gains of the 10th West Ridings and 11th West Yorkshires had significantly reduced the risk of enemy enfilade fire against the battalion's right flank.[3] Unfortunately, the same could not be said for its left flank.

The attack ran into trouble almost immediately. As the Green Howards climbed out into the open, they came under a devastating burst of heavy fire from the German machine gun on their left and quickly began taking losses.

Captain W.T. Wilkinson of C Company was among the first hit, cut down as he stood on the edge of the trench, encouraging his men forward.[4] The 29-year-old Londoner was the second company commander to die so far that day – the first, A Company's Captain F.A.H. Atkey, having been shot in the head by a sniper several hours earlier.[5] 'Great excitement prevails for we are to attack,' Atkey had written in his last letter home. 'The men are splendid and should do well.'[6]

Unfortunately, these same men were now in a perilous situation. Caught in the open, they desperately tried to escape the withering fire of the machine gun and rush across no-man's land toward their first objective, but few got far. Men were hit almost immediately, cut down mid-stride as the staccato rattle of the gun tore across the battlefield. Within minutes, the attack had faltered and was teetering on the brink of collapse.[7]

Not all had gone over with the initial assault. Second Lieutenant Donald Bell, now in command of the battalion bombers following the death of Lieutenant Gibson, was among those kept back, and he watched aghast from the relative safety of the British trenches as his comrades floundered just yards in front of him. It was clear to all that the attack had already reached a critical juncture and faced almost certain failure unless immediate action was taken.

At that moment, and without hesitation, Bell sprang into action. Armed with a supply of Mills bombs, he set off quickly down a communication trench, followed closely by two members of the bombing party: Corporal Harrison Colwill and Private Joseph Batey. The trench was littered with debris and the bodies of the dead, but the three men scrambled frantically along on their hands and knees, somehow avoiding detection.

When they got within twenty yards of the machine gun, Bell issued a quick order, drew breath and leapt from cover. With his Webley service revolver in one hand and a Mills bomb in the other, he charged across the open, straight at the startled Germans. 'The lieutenant got near enough to use his revolver,' recalled another British officer, watching the drama unfold with amazement. 'He knocked over the man working the gun. Then he flung a bomb at the outfit until it was smashed, and the whole of the men about it killed or wounded.'[8]

Remarkably, Bell later revealed he had only thrown one Mills bomb after dispatching the soldier manning the gun with his Webley. Still, it proved to be an exceptional throw, landing squarely on its target to knock out both the machine gun and the remaining members of the enemy gun team. This allowed the daring trio to race across the final few yards of open ground unmolested and down into the German trenches. There, they encountered several dugouts, which they swiftly began clearing with the assistance of further Mills bombs. Bell claimed this endeavour alone killed another fifty enemy soldiers.[9]

The extraordinary sequence of events lasted a matter of minutes, but they changed the course of the attack. With the threat on their left flank removed, the Green Howards stormed across no-man's land and entered Horseshoe Trench, where they quickly secured their objectives at Points 49 and 79. The brigade war diary notes that this renewed advance 'completed the demoralisation of the enemy', who surrendered in their droves. Those who did not capitulate, it added, were 'shot down and killed with the bayonet in considerable numbers'.[10]

By 7.00 pm, the entire length of Horseshoe Trench had been successfully occupied and cleared by the British, with the capture of 146 prisoners and two enemy machine guns.[11] Further advances that evening led to the eventual capture and consolidation of positions extending from Point 74 on the right flank to just west of Point 79 on the left. A detached post was also established and held at 'the Triangle', just over 300 yards to the east. During the night, German troops launched several counter-attacks against the left flank of the 9th Yorkshires in an effort to claw back some of the lost ground. These were all repelled, however, and by first light the situation remained unchanged.[12]

Enemy artillery activity remained constant throughout the following day, but no concerted attempt was made to retake Horseshoe Trench – much to the relief of the shattered Yorkshiremen. Instead, Bell and his men busied themselves consolidating their gains. They were assisted in this task by sappers from the 101st and 128th Field Companies RE, who moved forward to construct a series of strongpoints along the line held by the brigade. At the same time, new communication trenches were also dug across the old no-man's land by the 102nd Field Company RE.[13]

After nearly seventy-two hours under fire, the 9th Yorkshires' time in the front line came to an end during the evening of 6 July, when they were relieved by the 9th Welsh. On their right, the 8th Yorkshires handed over positions to elements of 68th Brigade. The relief was completed by 8.50 pm.[14]

The Green Howards had been pushed to the limits of their endurance but had fulfilled every task asked of them, fighting tenaciously with bayonet and bomb, in what was their first major offensive operation of the war. Despite their exhaustion, one observer noted a confident swagger to their step as they marched back to Belle Vue Farm, many festooned with souvenirs taken from the battlefield. Only when roll call was taken at 8.00 am the next morning did the true cost of their endeavours become apparent.

The attack on Horseshoe Trench was a tough introduction to large-scale offensive action, and initial casualty figures made for sobering reading. Three of the battalion's officers were killed during the assault, four more were wounded and two, initially reported missing, were later confirmed dead.[15] Also among the fallen was the battalion's respected medical officer, Captain John Cecil Rix, who was killed by a shell with two others as he tended the wounded outside his aid post. 'The two assistant orderlies and myself feel it all more than I can say,' his orderly wrote later. 'He was one of the happiest and most respected officers in the Regiment.'[16]

From the ranks, fourteen men were initially reported to have been killed and 144 wounded. A further twenty-eight were also missing. A more complete picture emerges after scouring the records held by the Commonwealth War Graves Commission. These show that forty-two members of the 9th Yorkshire Regiment lost their lives on 5 July – five officers and thirty-seven other ranks. They were the heaviest losses in the brigade and would be keenly felt in the days ahead.[17]

Nevertheless, the battalion had performed admirably, overcoming resolute enemy resistance to secure objectives of considerable difficulty. Their efforts did not go unnoticed at divisional headquarters. On 7 July, Major General J.M. Babington, the officer commanding the 23rd Division, addressed the battalion:

> 'The Division is proud of you. It is unnecessary for me to say I am proud of you – I always have been. You have made a great name for yourselves, your Division, and your Brigade. I don't wish to say more, but again, I thank you.'[18]

The battalion and its commanding officer also received formal commendation from Brigadier General T.S. Lambert in his report on the operations conducted between 3 and 5 July:

> 'The assault across the open by the 9th Yorkshire Regiment was most gallantly carried out in the face of heavy fire and this proved too much for the enemy's morale… I cannot speak too highly of the arrangements made by Lieutenant Colonel HOLMES, Commanding 9th Yorkshire Regiment in carrying out this assault which was completely successful. The losses in all Battalions were considerable but the energy in hunting out and destroying the enemy at the end of the long and confused operations in spite of their own physical exhaustion, was worthy of the highest traditions of the Yorkshire Regiments which they represented.'[19]

The success of the battalion's assault proved instrumental to the eventual capture of Horseshoe Trench by 69th Brigade, breaking the resolve of the enemy, who either fled or surrendered in significant numbers. Yet the attack had been close to collapse until Bell set off to destroy the machine gun with the equally courageous Corporal Colwill and Private Batey in tow. Their daring intervention averted near-certain disaster and wrested back the momentum of the attack. It also saved innumerable British lives.

Don was characteristically reticent about the entire episode. Writing to his mother from the relative calm of Belle Vue Farm on 7 July, he offered only a modest account of what had occurred:

> 'Dear Mother
> You will be surprised to hear from me again so soon, but as the other one I wrote earlier this morning was only a note & inherent at that, I will have another shot. As I told you the battalion has been in action & did splendidly, capturing a strong German position. I did not go over, as I was second in command of the bombers. Unfortunately, Gibson was knocked out so I carried on. As a result of Gibson's encounter, a machine gun was spotted on the left which could enfilade the whole of our front. When the battalion went over I with my team crawled up a communication trench and attacked the gun & the trench & I hit the gun first shot from about 20 yards and knocked it over. We then bombed the dugouts & did in about 50 Bosches [*sic*]. The G.O.C. [General Officer Commanding] has been over to congratulate the battalion

& he personally thanked me. I must confess that it was the biggest fluke alive and I did nothing. I confess I only chucked one bomb but it did the trick. The C.O. says I saved the situation for this gun was doing all the damage. I had a grand little lad with me, only 19, who did all the work & I think both he & I will be recommended for something, M.C. for me, & D.C.M. for him. I am glad I have been so fortunate for Pa's sake for I know he likes his lads to be at the top of the tree. He used to be always on about too much play and too little work, but my athletics came in handy this trip. We are out of [the] trenches at present and I am perfectly fit. The only thing is I am sore at elbows and knees with crawling over limestone flints. I haven't heard from you as yet this week but am hoping for better luck today. I hope you are all keeping well and remember don't worry about me [as] I believe that God is watching over me & it rests with him whether I pull through or not. I will write again as soon as I get another chance & will send Field P.C's every day if possible so that you will know that I am alright.

Goodbye for the present,
Love to all,
Don[20]

With some rare free time that afternoon, he also wrote to his sister Nancy and was equally self-effacing about the incident.

> 'Since I last wrote I have been living in stirring times for we have been taking part in the great offensive of which you will have already heard. The Battalion went over and did splendidly, capturing a very strong position. I did not go over with them but did a bombing stint on the left and was lucky enough to knock out a machine gun which was causing the lads some bother. The CO was very pleased and I hear I am to be recommended for something – shall have to wait and see. It really was the biggest fluke in the world and I did not think anything about it. Of course I could not see what was happening but I understand that as soon as the machine gun stopped our fellows came across. We are out of the trenches at present having a well-earned rest.'[21]

Nancy was now 6,000 miles away in South Africa, having left Britain in July 1915 to take up a post as 'Music and Elocution Mistress'[22] at the Wesley Girls' School in Grahamstown, Eastern Cape Province. She had also recently married Harry Sole, a local printer, at the town's Wesleyan Methodist Church. Don maintained correspondence with his sister, usually through letters sent via Milton Lodge, and was always eager to hear about her new life. Nancy had only just turned 21 when she set sail for Port Elizabeth on board a Union-Castle liner, accompanied by nothing but a black wooden trunk holding her belongings. Yet she shared the same streak of determination and quiet resolve that characterized each of the Bell children, particularly Don.

Her youngest brother may have sought to downplay his role in the attack at Horseshoe Trench, but his actions had not escaped notice at brigade headquarters. Brigadier General Lambert made direct reference to him in his after-action report and was unequivocal in his belief that the young officer deserved proper recognition for his deeds. 'There were many actions of individual heroism,' noted Lambert, 'among which that of 2nd Lieutenant D.S. Bell, 9th Yorkshire Regiment was worthy, in my opinion, of the highest award.'[23]

News of Bell's exploits on the Somme soon reached his hometown, where the local press were quick to heap praise. The editor of the *Harrogate Herald*, William H. Breare, maintained correspondence with many local soldiers and was among those most eager to share details of the heroics in one of his weekly columns:

> 'Second Lieutenant Don Bell was in charge of the bombing section. The boys who were to make the rush went on. Soon a German gun was discovered somewhere on their flank enfilading their ranks and doing much damage. Don Bell took some bombers in the direction of this gun. They crawled on their hands and knees ever so far until within about twenty yards of the offending gun. Bell threw one bomb, and in that first shot blew the gun to smithereens. The party then stormed the German trench and sent fifty Germans below. Next day the General Officer Commanding came to the regiment and thanked them for the success of that great movement, for it had been entirely successful. He personally thanked Second Lieutenant Don Bell, and the Commanding Officer declared that he had saved the situation.'[24]

In response to Bell's modest dismissal of the assault's success as nothing more than a 'fluke', Breare had this to say:

> 'A fluke indeed! We have only to inquire what a fluke it is to know how far wrong Don is in his estimate. What is a fluke? A fluke is the accomplishment, by an unskilled person, of something for which he had not tried. In other words,

> an occurrence due to accident rather than skill. Now, too many of us know Don Bell as one of our finest athletes. We know what he has accomplished in the world of sport. We know that he could always throw straight. He could not deny that he tried to hit the gun, or that he did it in the first throw. He is not an unskilled person. Yes, he did try for his objective, and what is more, succeeded... Undoubtedly Don saved the situation. That is why we are nearly bursting with pride.'[25]

By his own admission, Bell had been extraordinarily fortunate to come through the attack unscathed. It was remarkable that neither he, Colwill nor Batey were hit as they dashed across the open ground in full view of the enemy. The sheer audacity of the assault likely contributed to its success, catching the German machine-gun team completely off guard. However, the undeniable element of good fortune should not detract from the extreme bravery displayed by the three men, particularly Don, whose speed and guile would have been familiar to anyone who had seen him on the football pitch.

A few hours after writing to his mother and sister, Bell and the 9th Yorkshires were ordered to positions near Bécourt Wood, where they were to be held in reserve as 24th Brigade attacked Contalmaison. The 10th West Ridings and 11th West Yorkshires had also been placed at the disposal of that brigade earlier in the day.[26] It was now raining incessantly, and most of the battalion were already soaked by the time they trudged out of Belle Vue Farm at 7.30 pm, making the 1½-mile march a miserable one indeed.[27]

The atrocious conditions were also proving of significant hinderance to the troops attempting to capture the village, and a steady stream of casualties were reported throughout the day as the attacking infantry struggled through the mud. Nevertheless, elements of the 1st Worcesters did succeed in fighting their way into the village, holding it briefly before being forced to withdraw under intense enemy artillery fire. 'All during the operations of the 7th, rain fell almost continuously,' noted 24th Brigade's war diary, 'and the state of the trenches was appalling, owing to the mud and congestion by wounded and dead of both this Brigade and the 17th Division.'[28]

Thankfully for Bell and his weary comrades, the battalion was not called upon, and at 11.00 pm they made their way back to Belle Vue Farm – cold and wet, but relieved to be out of range of all but the heaviest German guns.[29]

They were joined there the next morning by the 11th West Yorkshires, but the men of the 8th Yorkshires were not so fortunate. They remained in positions immediately in front of Bécourt Wood for a further two days, losing several men to enemy shellfire.[30] One heavy shell even collapsed the dugout housing battalion headquarters, though miraculously, its occupants all emerged from the rubble unscathed. The constant bark of nearby British batteries also made sleep difficult, and many in the battalion were close to exhaustion when they finally returned to Belle Vue Farm on the evening of 9 July.

Unfortunately, less than twelve hours later, both battalions of Green Howards would be heading back toward the front, where they would spearhead the next British attempt to take Contalmaison.

Chapter 11

A Tiny French Hamlet

Contalmaison lies 1½ miles south of the old Roman road that cuts gun-barrel straight through the battlefield between Albert and Bapaume. Strategically positioned on a spur of the Pozières Ridge, the village provides commanding views over much of the surrounding countryside and was once notable for a majestic château that stood in spacious grounds at its northern extremity. To the east lies the brooding mass of Mametz Wood, where the ground rises before falling away gently towards Bazentin-le-Petit. To the west, the village is flanked by three much smaller wooded areas – Contalmaison Wood, Bailiff Wood and Peake Wood – skirting some 300 yards beyond its boundary, on a north–south axis. It is a typically sleepy rural hamlet with a church at its centre and a sedate way of life. It was much the same in the years leading up to the First World War, when 161 people lived in the village, though in those days it could boast its own café, school and even a tennis court.[1]

The author and future poet laureate, John Masefield, spent time in Contalmaison during the war, and later wrote about the village in his book *The Battle of the Somme*:

> 'In itself, it was a tiny French hamlet at a point where a road from Fricourt to Pozières crosses a road from La Boisselle to Bazentin. It may have contained as many as fifty families in the old days before the war. Most of these were occupied on the land, but there was also a local industry, done by women, children, and old men, of the making of pearl-buttons. There was a church in the heart of the village, and just to the north of it a big three-storied French château, in red brick, with white and yellow facings, and a turret en poivrière in the modern style. This château stood slightly above the rest of the village.'[2]

Fighting first broke out north-west of Contalmaison in mid-September 1914, and the village was eventually captured by German troops from

Badisches Reserve-Infanterie-Regiment (RIR) Nr.109, supported by Württembergisches RIR Nr.120, on 28 September.[3] A series of costly French counter-attacks followed but achieved little. By the end of the year, the front line had stabilized around Ovillers and La Boisselle to the west, and Fricourt and Mametz to the south. Contalmaison now sat almost two miles behind the front and became a quiet backwater, positioned between the first and second German defensive lines.

The village remained largely undisturbed for the next eighteen months, its buildings spared significant damage despite a noticeable upturn in artillery activity when the British began extending their responsibilities southward from August 1915.[4]

By the summer of 1916, Contalmaison had become an integral part of the German defensive system and a veritable fortress.[5] It was protected by trenches and thick belts of barbed wire, while carefully positioned machine-gun emplacements commanded clear fields of fire over the approaches from the south and west.[6] Below ground, cellars had been fortified to provide shelter from all but the biggest of shells. Its strategic importance was fully understood by the British, who knew its capture was critical for any operations in the area.[7]

On 1 July 1916, the village formed part of the third line of objectives. 'This line, to be reached by the 103rd Brigade at 10.10 A.M.,' wrote Captain Miles in the official history, 'was to be put into a state of defence preparatory to a subsequent assault on the German 2nd Position, which lay eight hundred yards beyond.'[8] The second line, including the northernmost section of Quadrangle Trench, which protected the village's western flank, was 'to be reached by 8.58 A.M., when the 101st and 102nd Brigades were to halt and consolidate'.[9]

With zero hour fixed at 7.30 am, the timings laid down were wildly optimistic, even if things went according to plan. Unfortunately, the attack by Ingouville-Williams's 34th Division astride the main road at La Boisselle proved to be an almost unmitigated disaster. Wave after wave of infantry were cut down before a tenuous foothold was secured in and around the crater at *Schwabenhöhe* and the German line further south.[10] 'The serried ranks of the enemy were only a few metres away from the trenches when they were sprayed with a hurricane of defensive fire,' recalled one German soldier. 'Our weapons fired away ceaselessly for two hours, then the battle died away.'[11]

Against all odds, a small party of men from the 16th Royal Scots,[12] and several others from the 27th and 24th Northumberland Fusiliers, did advance as far as Contalmaison on the opening day of the offensive. They were too few in number to make any real difference, however, and those not killed or taken prisoner were forced to withdraw after some brutal close-quarters fighting.[13] A handful of wounded men were captured and remained in the village until it finally fell nine days later. 'There were eight or nine other Englishmen, all wounded, lying there,' said one later. 'I was in front, right in the mouth of the dug-out, where I could see the trench, where a lot o' Boches was sitting, smoking cigarettes, an' talking in their own lingo.'[14] In reality, few British soldiers got anywhere near Contalmaison that day. Only limited gains were achieved south of La Boisselle, including the capture of Sausage Redoubt and Scots Redoubt.[15] It was meagre return for the amount of blood spilt, serving as a grim prelude to the ferocious fighting that lay ahead.

The ninth of July dawned with all the promise of a fine summer's day. The wet weather that had dogged British operations in recent days had finally lifted, bringing clear but slightly cooler conditions that 'generally favoured offensive operations'.[16] At Belle Vue Farm, Bell and his men spent much of the day cleaning kit and equipment, their attention occasionally drawn to the steady drone of aircraft passing overhead.[17] With reduced cloud cover, the Royal Flying Corps were at last able to provide full air support for the British batteries, and much-needed target registration and aerial observation had been stepped up over the sector. Bombing sorties also intensified, with machines dispatched to target enemy railheads, supply dumps, bivouacs and headquarters as far back as Cambrai, some 25 miles behind German lines.[18]

After the failed attempt to take Contalmaison on 7 July, a renewed effort to push into the village from the south had been attempted the next day. Unfortunately, this was stopped abruptly by concentrated machine-gun and artillery fire and eventually abandoned. 'Conditions were appalling,' records the history of the 23rd Division. 'The flooded trenches, where men could scarcely drag along, were now congested with killed and wounded. Nevertheless, continuous efforts were made to establish posts at the southern end of Contalmaison. But every fresh attempt failed under heavy hostile fire.'[19]

Responsibility for capturing the village now passed to 69th Brigade. At 11.00 pm, orders were issued for a fresh assault to be launched on the afternoon of 10 July. The attack would be undertaken by the 8th and 9th Yorkshires, advancing from the west along a frontage of just under 800 yards. Simultaneously, the 11th West Yorkshires would advance through Bailiff Wood to secure the left flank. Zero hour was initially set for 4.30 pm, though this would be pushed back to 4.50 pm, with a preparatory bombardment and artillery barrage on the village – as well as targets to the north and east – preceding the attack.[20] Machine guns and trench mortar batteries were tasked with providing additional supporting fire. To ensure the assault proceeded in a coordinated manner, Bell and his fellow officers, and the brigade's non-commissioned officers, were to conduct reconnaissance and study aerial photographs and intelligence reports provided by army headquarters. It was hoped this would enable every man to reach his designated objective and minimize any delay in the consolidation of those positions once secured.[21]

At 11.00 am the next day, the 9th Yorkshires trudged out of Belle Vue Farm towards the front. The five-mile route forward proved difficult, as ground conditions were poor and the roads heavily congested. The battalion was also harassed by German artillery fire along the way, causing further delays. By the time they finally reached their assembly trenches at 4.00 pm, the Green Howards were already thirty minutes behind schedule.[22] Worse still, they found the positions so badly damaged that, in places, they were 'barely trenches at all'.

With time against them, alternative positions were quickly identified and three companies were dispatched to occupy a section of the *Frohlich-Graben* between Point 81 in the north and Lincoln Redoubt in the south. The remaining company took up position approximately 250 yards to the rear, along the *Steinmann-Weg*.[23] To their right, the assembly trenches of the 8th Yorkshires were also in poor condition. However, after quick deliberation, they were considered suitable and occupied as originally instructed.[24]

No sooner had Bell and his battalion taken up their positions than the ear-splitting boom of the British artillery reverberated through the air, signalling the start of the bombardment. Within minutes, Contalmaison had all but disappeared from view as shells poured into the village and positions to the north and east. For those British troops with a vantage point, it was an exhilarating spectacle.[25] Watching dense plumes of black smoke and red brick dust belch out of the ruins and spiral into the sky, it was difficult to imagine anyone – or anything – surviving such a brutal onslaught. Some of the Green Howards even felt a degree of pity for those unfortunate enough to be caught underneath it.

Across the shallow valley, the German garrison – troops from I Battalion, RIR 212 – were indeed suffering. Although a renewed British attempt to dislodge them from the village had been expected, the timing and sheer intensity of the bombardment was already taking a heavy toll. Some enemy soldiers tried to flee, but the majority remained to defend their positions, and these now faced a hellish struggle for survival. One later described the devastating effects of the shellfire as it pulverized everything around them:

> 'So heavy was the English bombardment that our positions at the edge of the village were soon buried. Above the ruins, shrouded by smoke and fumes, English planes circled overhead. One dug-out after another collapsed, one machine gun after another destroyed, the cellars of the château were already full of wounded men… Scarcely more than a hundred escaped to our second position.'[26]

Well aware that a fresh attack was imminent, the Germans made several futile attempts to secure their right flank before the main infantry assault began. The first came at 4.10 pm, when approximately 200 enemy soldiers launched a counter-attack toward Points 46 and 55 near Bailiff Wood.[27] Advancing from the north, they were met by concentrated British machine-gun fire and quickly driven back towards Contalmaison Wood. Forty minutes later, a smaller force of thirty to forty men attempted to bomb their way west along the *Kaiser-Graben* from the north-western corner of the village, but they too were swiftly repelled, this time by Lewis gun and rifle fire.[28]

Those caught under the bombardment could do little but seek refuge in the few remaining cellars or deep dugouts still intact. Conditions inside were appalling: dead and wounded men lay everywhere, and the air was fouled by the stench of sweat, blood and human waste. Some shelters were so overcrowded that men were forced to wait outside in the bombardment before taking their turn inside. This had predictably catastrophic consequences, as one German soldier recalled:

> 'Those who went outside were killed or wounded... Some of them had their heads blown off, and some of them had both their legs torn off, and some of them their arms. But we went on taking turns in the hole, although those who went outside knew that it was their turn to die, most likely. At last most of those who came into the hole were wounded, some of them badly, so that we lay in blood.'[29]

Some enemy soldiers did manage to escape the village, but few got very far. At 4.30 pm, around 200 were observed attempting to flee north toward Pozières, but these quickly came under heavy British machine-gun fire and were cut down in large numbers. More were seen running east in the direction of Bazentin soon after, but these met a similar fate.[30] It was a desperate situation. After enduring two days and nights of relentless shellfire, German morale was plummeting. To compound their misery, food supplies were also low as ration parties were unable to get through to the village from the rear, and fresh water was becoming increasingly scarce.[31]

Of greater concern to the German command was the fact that the garrison had already been reduced to fewer than 600 men of all ranks, seriously undermining its ability to mount an effective defence when the time came to venture out and face the British.[32] Small detachments of men had been rushed forward, but they were too few in number to make any meaningful difference.[33] Under such trying conditions, most German soldiers were reduced to cowering together in their overcrowded dugouts, half-deafened by shellfire and near suffocated by fumes, as they waited anxiously for the bombardment to lift and the attack to begin.[34]

The sight of Contalmaison being pounded by the British guns gave the watching Yorkshiremen a renewed sense of confidence, but few underestimated the difficulty of the task at hand.[35] They had been well and truly blooded at Horseshoe Trench, and many men were haunted by thoughts of death, or perhaps more dreadful, of horrific injury and grotesque disfigurement. The burden was greater still on those leading them into battle. Junior officers knew they were statistically far more likely to be killed or wounded than those under their command,[36] yet the vast majority, like Donald Bell, fulfilled their duty with extraordinary grit and self-sacrifice. For many young subalterns, the greatest fear was not of death itself, but of failing in front of their men.

As the final seconds ticked slowly down, thick smoke 'belched over the battlefield, and the song of the shells was loud and high'.[37] Steeling themselves for the immense psychological effort required to haul their bodies back out into battle, few of the Green Howards even noticed when the howitzer batteries targeting the village and its western defences stepped down their rate of fire slightly at 4.50 pm.[38] But then the whistles blew and in unison, along the 600-yard front, an extended line of khaki-clad soldiers clambered up from the dirt into the vast, open expanse of no-man's land to begin the attack.[39]

Chapter 12

Into the Breach

The advance on Contalmaison would be undertaken in four waves per company, with searching and consolidating parties following in similar formation. To reduce the time spent in the open, the Green Howards were ordered to move in quick time.[1] Even so, much of the attack would take place under direct German observation, with those troops assembled furthest back at the *Frohlich-Graben* forced to cover more than 1,500 yards of mostly exposed ground before reaching their objectives.

Before they could even move off, each of the 9th Battalion's four companies had to first form up in the open to correct their alignment. This manoeuvre, necessitated by the late change to their assembly positions, might easily have proven problematic, had it not been for the 'admirable preparations' put in place by Lieutenant Colonel Holmes and Lieutenant Colonel Vaughan, whose sterling work ensured 'no confusion or misdirection occurred'.[2] With the task swiftly accomplished, the attackers – bayonets glinting in the warm afternoon sun – finally set off in extended order towards the village and their first objective.

Bell and his men were initially greeted by an eerie stillness that caught many by surprise. Some soldiers even recalled hearing the faint sound of birdsong through the rapport of the guns, though the quietude was fleeting.[3] When they crested *Jäger Höhe*, bringing them into full view of the village, albeit partially obscured by thick dust and smoke, a cluster of red signal flares could be seen arcing skyward, 'fired in profusion by the German infantry calling for artillery support'.[4] The response was swift and brutal. Moments later, the Yorkshiremen came under heavy and accurate artillery fire from the front and both flanks. At the same time, enemy machine guns in the village and Pearl Alley to the south opened up, traversing along the advancing lines of British soldiers. Caught in this cauldron of high explosive, shrapnel, and machine-gun fire, gaps soon began to appear in the ranks of the Green Howards, yet each man pressed on 'steadily and without hesitation'.[5]

In an effort to cover the advance, a smoke barrage had been put down by four 4in Stokes guns positioned around 400 yards west of Bailiff Wood.[6] Unfortunately, this was largely ineffectual, as too few bombs could be brought forward in time for the attack. To the south, two Stokes guns ably supported the right flank from positions near the northern end of Shelter Alley and were later hauled into the village. No crews were available to operate the remaining two guns, but their ammunition was redistributed to resupply the others.[7]

British machine-gun support proved markedly more effective. Before the attack, the 10th West Ridings had established posts south of the village, from where a Vickers gun now swept broad sections of the battlefield. Additional guns were positioned at points to provide enfilade fire across the village and along both flanks, while to the rear, yet more guns maintained indirect fire on likely approach routes from the south and south-east. Throughout the assault, these proved invaluable, inflicting significant losses on the enemy and disrupting attempts to reinforce or counter-attack.[8]

By now, Contalmaison lay in ruins. Two days and nights of relentless shelling had reduced the village to rubble, and brick dust hung thick in the air. The incessant shellfire, repeated alarms and constant need to repair their damaged trenches and dugouts had exhausted the German defenders.[9] According to the regimental diary, they were so demoralized that only one officer and fifteen men from the two companies on the right emerged to face the assault when the alarm sounded at 4.50 pm.[10] These moved quickly to take up position among the ruins at the western edge of the village, but visibility was so poor they could scarcely see the British troops coming towards them.[11]

On the left of the attack, Bell and the 9th Yorkshires were making steady progress. After dropping down from the modest heights of *Jäger Höhe*, they crossed the light railway that ran through the valley, before sweeping up the gentle rise towards the village.[12] Within no time, they were bearing down on the first major obstacle in their path: Quadrangle Trench, which sat around 250 yards west of the village on a north–south axis. Aerial reconnaissance had identified thick banks of barbed wire in front of the trench, which was believed to be particularly well held on the south-western outskirts, where a strongpoint known as Leith Fort was situated.[13]

The area had been torn apart by the British guns and was 'broken by innumerable shell holes'.[14] Yet there was still a great deal of apprehension in the ranks of the approaching Green Howards. As they moved in with fixed bayonets, they could see enemy soldiers desperately attempting to retreat back through the barrage and into the ruins of the village.[15] The effectiveness of the bombardment here was immediately apparent. Mangled heaps of wire and twisted metal lay strewn across the ground, while the bodies of dead and dying enemy soldiers – many dismembered or grotesquely mutilated – lay partially buried, almost indistinguishable from the smouldering wreckage.

While some Germans 'still fought stoutly',[16] Bell and his fired-up comrades quickly surged across the grizzly stretch of land. By 5.00 pm, the northern portion of Quadrangle Trench had been cleared and secured, with relatively modest casualties.[17] Their advance had benefited considerably from the actions of the two companies of the 11th West Yorkshires on their left, which, after occupying Bailiff Wood – a newly planted spinney of saplings only a few feet high[18] – had pushed east along the *Kaiser-Graben* towards the north-western corner of the village. This helped protect the left flank of the attack, greatly diminishing the threat of enemy enfilade fire from the north.[19]

Unfortunately, on their right, the 8th Yorkshires had encountered significantly heavier fire as they descended into the valley, sustaining considerably more casualties as a consequence. This was due in large part to the greater concentration of enemy machine guns positioned along the southern edge of the village, including several covering the south-eastern approaches – sited there in anticipation of another assault from that direction.[20] These guns proved to be of particular menace to the right flank of the attack, cutting through the exposed lines of infantry on the open slope with devastating efficiency.[21]

When the 8th Yorkshires finally reached Quadrangle Trench at 5.25 pm, they were confronted by large swathes of uncut wire and a quagmire of water-filled shell holes. The ground was so badly churned up that the only practical means of advancing was in small groups, moving forward in short rushes of only a few yards at a time.[22] This left them extremely vulnerable to enemy fire as they tried to pick their way through the German defences.

Even after securing the trench, the hard-pressed infantry found little cover, as it was only 2–3ft deep. In the confusion, some men also ran into their own barrage, prompting frantic signals for the lifts to be adjusted. These were quickly implemented, but not before further losses were incurred.[23]

With both battalions now occupying Quadrangle Trench, the Green Howards prepared to resume their attack. Many were already fatigued from the long approach and parched from the stifling heat. Nevertheless, when the barrage lifted at 5.30 pm, they rose from their positions and dashed into the village.[24] On the left, Bell and his battalion quickly pushed through the ruins, killing or capturing any enemy soldier offering resistance. When they reached the *Rue Vicaire*, they wheeled left and ensconced themselves in an enemy trench running along the northern perimeter. Behind them, search parties moved in to begin the mopping-up process, during which around 100 Germans and two machine guns were captured.[25] More prisoners – many seriously wounded – were taken from dugouts found in the rubble. It was clear they had suffered tremendously during the bombardment.[26]

At the southern end of Contalmaison, the 8th Battalion's advance had been stalled by the unexpected discovery of a garden hedge, reinforced with wire netting, holding the line just outside the village. As groups of men attempted to negotiate this improvised barrier, they came under heavy fire from some Germans holed up in ruins on their left.[27] These were eventually subdued and the hedge traversed, but not without loss. Indeed, it was later reported that half of the battalion's casualties that day occurred between Quadrangle Trench and the hedge, a distance of just fifty yards.[28]

The depleted but resolute ranks of the 8th Yorkshires pressed on into the village. Ahead of them, they could see enemy troops fleeing east and set off in quick pursuit. German resolve was now completely broken. In no time, the final pockets of resistance had been overrun, and those Germans not killed or captured had been driven back towards the Cutting and Pearl Alley.[29] With the southern portion of Contalmaison secured, mopping-up parties began combing the ruins for survivors. The official history notes the battalion, 'reduced to five officers and 150 men, had the satisfaction of taking prisoner a force greater than its own: eight officers and 180 unwounded men of the 122nd Reserve Regiment'.[30] Among their number was the garrison commander, who was found in the cellar of the château.[31] A further 100 wounded men were discovered sheltering in various dugouts elsewhere. Many of them seemed resigned to their fate after the brutal bombardment:

> 'We stayed down in the hole waiting for the end. Then we heard your soldiers shouting. Presently two of them came down into our hole. They were two boys and they had their pockets full of bombs... They had bombs in their hands also, and they seemed to wonder whether they would kill us. But we were all wounded, nearly all, and we cried "Kameraden!"... And now we are prisoners – and I am thirsty.'[32]

The condition of some of these prisoners was pitiful. 'The men left in Contalmaison were in a dreadful state, having suffered to the very limit of human endurance, and beyond,' one British observer wrote later. 'Soon the rations they had brought with them were finished, and owing to our ceaseless gun-fire it was impossible to get fresh supplies. They suffered great agonies of thirst, and the numbers of their dead and wounded increased steadily.'[33]

The 8th Yorkshires also captured six enemy machine guns and several thousand rounds of ammunition during the assault. It came at some cost, however, with a force barely 150 strong left fighting alongside their commanding officer, Lieutenant Colonel Vaughan.[34] Most of the wounded were already streaming back to advanced dressing stations behind the lines, where they were met by members of the Royal Army Medical Corps before being patched up or sent on to Casualty Clearing Stations:

> 'Presently the wounded began to pour in, some on stretchers, others more lightly wounded, walking. Each officer at the dressing-station had a bell-tent where he examined and treated each case in turn. Hundreds of wounded men were soon gathered in the open space, some sitting, others lying on the ground. While they waited, the orderlies supplied them with cups of hot tea or cocoa, and pieces of bread or biscuit and bully beef. The "walkers" were then sent off in cheerful parties of fifty or more to follow the sign boards which pointed the way to the next dressing-station further back, while the stretcher cases were put on to motor ambulances, or empty ammunition lorries, which hurried as quickly as possible out of the danger zone. So great was the rush of wounded that kept pouring down from Peake Wood and Bailiff Wood that we soon had to give up our tents and attend to them on the open ground outside. Before long the whole place looked like a shambles, with wounded

> and bandaged men lying everywhere, and it was a most fortunate thing for us all that the enemy was too much engaged elsewhere to have time to shell the areas further back.'[35]

The shattered remains of Contalmaison had been occupied, but much remained to be done before the British could claim its successful capture. It was now a race against time, with the weary Yorkshiremen moving swiftly to consolidate their tenuous hold on the village before the inevitable enemy counter-attacks began. A line of defence was established, skirting along the northern perimeter before swinging south-east, through the vicinity of the ruined château and the *Vieux Manoir* ('Old Manor').[36] From there, men from the 8th Battalion were ordered to dig a new trench down towards Point 84, close to the junction with the *Rue de Mametz*. It had initially been intended to consolidate the entire outer perimeter of the village, but this shorter line was now deemed more appropriate given the limited number of men available.[37]

At the northern end of Contalmaison, the 9th Yorkshires had linked up with the 11th West Yorkshires on their left, establishing a strongpoint at Point 25 with a trench block and two Lewis guns. These provided vital covering fire across the northern approaches in the event of any counter-attack from that direction.[38] The 9th Battalion's right flank extended to the eastern wall of the château, beyond which men from the 8th Battalion had already started work on the new trench leading south. They were later assisted in this task by two reserve companies of the 11th West Yorkshires and soldiers from the 10th West Ridings, but these reinforcements did not arrive until after 11.00 pm.[39]

The first indication of enemy activity came at approximately 7.30 pm, when a small group of Germans emerged from positions close to the Cutting.[40] These were quickly dispersed with the aid of a captured machine gun and did not reappear. Further south, the situation was more precarious. Troops here were too few in number to mount a defence of any considerable strength, which left the positions held by men from the 8th Battalion dangerously exposed and susceptible to counter-attack until reinforcements could be brought forward.[41]

These vulnerabilities threatened to be exploited at 9.00 pm, when a group of around forty Germans were spotted taking cover along a hedgerow extending north-east from the *Rue de Mametz* at Point 93, less than 80 yards from the thinly held British line. Moments later, a sudden and sustained burst of fire was directed towards the Yorkshiremen, and enemy soldiers were observed making efforts to outflank their position from the south.[42] The threat this posed was immediately obvious, as any breakthrough here would leave the defences in the village open to attack from the rear. The situation, it was noted, 'gave rise to some anxiety'.[43]

In response, the battalion's second in command, Major Western, and several of his men hauled a makeshift barricade across the road, despite being under heavy fire. This succeeded in temporarily checking the German attack, buying critical time for runners to be quickly dispatched across the village.

At the other end of Contalmaison, the men of the 9th Battalion were still labouring to consolidate their positions when the sound of rifle fire broke out to the south. Moments later, a breathless runner arrived with a desperate plea for reinforcements, confirming – if there was any lingering doubt – that things there were indeed serious. It prompted a swift response from Lieutenant Colonel Holmes, commanding the 9th Yorkshires.[44] Gathering what few men he could spare, including Second Lieutenant Donald Bell and several bombers, Holmes sent them off through the ruined streets to assist. At the same time, a small number of 8th Battalion men were also hurrying to the aid of their beleaguered comrades.[45]

At the south-eastern perimeter, the fighting was both chaotic and deteriorating by the minute. Major Western's hastily constructed barricade had thus far frustrated German attempts to outflank their positions, but things were becoming increasingly precarious. With dusk closing in and enemy troops pressing ever closer, it seemed inevitable that they would soon overwhelm the exhausted Yorkshiremen and snatch back a crucial foothold in the village.[46]

Bell was among the first reinforcements on the scene and took immediate action. After assembling a handful of bombers, he issued orders for an imminent attack. Moments later, and with scant regard for his own safety, he sprang from cover and charged towards the enemy, hurling bombs with one hand while firing his Webley with the other. Within seconds, he was hit – a bullet pierced the front of his steel helmet, grazed his scalp and exited through the rear. It was a miraculous escape, but blood soon poured profusely from the wound, streaming down into his eyes and across his face.[47]

Driven by adrenaline and sheer force of will, he pressed on. A ferocious close-quarters bomb fight now took place, illuminating the evening sky with each flash of exploding grenade. 'He dashed forward with an armful of bombs and started to clear out a hornet's nest of Huns who were ready to take toll of our advancing troops,' recalled one astonished onlooker later. 'He advanced with great courage right up to where the enemy were posted. He took careful aim and bowled out several of the Germans.'[48] Bell was struck several more times but again escaped serious injury, even though one piece of shrapnel took a chunk from his helmet and ripped through the front of his tunic.

After a relatively brief but violent exchange, the 'extreme gallantry' of the British bombers prevailed, forcing the Germans to turn tail and fall back towards Pearl Alley under a flurry of Mills bombs. As they withdrew, Bell's luck finally deserted him: he was struck by a large fragment of shell that tore straight through his helmet and penetrated his left shoulder. The jagged piece of metal inflicted catastrophic injuries, causing him to stagger a few yards before collapsing to the ground. An eyewitness later described the dramatic last moments of the fearless young officer's life:

> 'Unfortunately, he was hit. For a while he fought on, but he was hit again. He got weaker and had to relax his efforts. He collapsed suddenly. When we reached him he was dead. Every officer and man of the company felt they had lost a personal friend, and his death was greatly deplored, though grief was qualified by pride in the fact that he had met the death he would have wished.'[49]

The enemy attack was abandoned entirely shortly afterwards when a Lewis gun was brought into action from the ruins of a nearby building. British batteries were thereafter able to maintain a continuous barrage on the enemy trenches, which proved sufficient to prevent any further counter-attacks.[50]

Though ultimately unsuccessful, German efforts to break through the thinly held British defences laid bare just how precarious the hold on Contalmaison remained until reinforcements could be brought forward.

Overstretched and outnumbered, the Green Howards had come alarmingly close to losing control of the situation – an outcome that would have jeopardized the entire operation. That they held firm is a testament to the leadership of Major Western, subsequently awarded the Distinguished Service Order, and to the steadfast resolve of the men under his command.[51]

However, it was the timely intervention of Second Lieutenant Donald Bell and his party of bombers that ultimately proved decisive in repelling the enemy – a fact later acknowledged by Lieutenant Colonel Holmes. Writing to Smith Bell on 15 July 1916, Holmes told him that his son's gallant conduct had 'once again saved the situation'.[52] Indeed, the events of 10 July bore a striking resemblance to those at Horseshoe Trench five days earlier. On both occasions, Bell's quick thinking and courage under fire helped to seize back the initiative at a critical moment in the fighting. His actions saved countless lives and proved strategically decisive in the capture of both Horseshoe Trench and the village of Contalmaison.[53]

When it was safe to do so, Bell's body was recovered from the battlefield and buried close to where he had fallen. Two 8th Battalion men – Lance Sergeant J. Whiteley, MM, and Private J. Devlin – were buried nearby, as was Second Lieutenant J.D. Marks of the 10th West Ridings. All three were killed during the fighting that evening.[54]

The first reinforcements reached the village around 11.00 pm, when two companies of the 11th West Yorkshires, laden with much-needed ammunition and supplies, made their way through a heavy barrage from Scots Redoubt.[55] They were soon joined by elements of the 10th West Ridings, who had already endured a torrid forty-eight hours under fire. Both battalions would toil through the night with the surviving Green Howards to consolidate their positions in the village, despite being subjected to heavy shellfire. Several patrols were also sent out, but aside from a brief encounter with a small party of Germans near the Cutting, all returned without incident.[56]

By midday on 11 July, British troops were firmly entrenched in Contalmaison. Along the northern perimeter, the 9th Yorkshires had constructed several strongpoints supported by machine guns. Additional defensive positions were also established at the Old Manor and on the south-eastern edge of the village by the 8th Yorkshires, the 10th West Ridings and two companies of the 11th West Yorkshires.[57] All available machine guns and Lewis guns had also been brought forward and positioned at strategic points around the village. Two companies and a section of Royal Engineers had by now reached the village, after failing to get through the German barrage at dusk the previous day. These greatly assisted in further strengthening the British defences.

The 9th Yorkshires were finally relieved in the early hours of 12 July by the 1st Black Watch. They had endured a tumultuous week but had clearly proven their worth in battle. As the remnants of the battalion marched to the rear – filthy, unshaven and hollow-eyed – they passed official war correspondent Philip Gibbs, who later wrote:

> 'They were the men who captured Contalmaison the day before yesterday, and they were marching with such a steady swing that it was hard to think they had been through such fighting and fatigues, and that they had left behind the many good fellows who will never come back. Along the road they were bringing back trophies of victory. On their waggons beside their own steel hats were German helmets. Some of the enemy's machine-guns were passing back with them, and although the men were tired, they held their heads high, and there was a fine pride in their eyes. An officer who watched them pass called out the names of their regiments, and said "Well done", and one of their own officers waved his hand and called back, "Cheer O". It was the greeting of gallant fighting men.'[58]

The capture of Contalmaison was achieved through courage and perseverance, but it came at considerable cost. Brigadier General Lambert described the brigade's losses as 'not excessive considering the importance of the operations',[59] but did concede they were heavy. Colonel Wylly, writing in the regimental history, offered a blunt and far less ambiguous appraisal of the 9th Battalion's casualties:

> 'The price which had been paid for such decisive results was, however, a heavy one. When the roll was called in the village of Contalmaison on the night of 10th July, there were present besides the Commanding Officer, Lieut.-Colonel H.G. Holmes, and the Second-in-Command, Major H.A.S. Prior, only five subaltern officers and 128 other ranks, while of the seven remaining officers, two, Major Prior and Second-Lieutenant Bingham, had been wounded; in all three officers and 13 other ranks had been killed, 11 officers and 192 non-

> commissioned officers and men were wounded, while 24 men were reported missing, and in the week's fighting the 9th Battalion The Green Howards had sustained losses in killed, wounded and missing to the number of 23 officers and 415 other ranks. The names of the officers who were killed or died of wounds received on the 10th were Major V.L.S. Beckett, Lieutenant [*sic*] D.S. Bell and Second-Lieutenant T.T. Wood.'[60]

In total, three of the battalion's officers and fifty-one other ranks were killed during the operation to take the village.[61] While these losses were heavy – exceeding those sustained at Horseshoe Trench five days earlier – they were marginally fewer than those of their sister battalion. The war diary of the 8th Yorkshires initially recorded the deaths of six officers and nineteen other ranks, with a further six officers and 231 men wounded, and twenty-seven reported missing.[62] However, revised figures later confirmed that six officers and sixty-nine other ranks had perished.[63]

It would take time for the Green Howards to recover from their ordeal. The loss of so many original members of the battalion rendered the success bittersweet for many, particularly those who had lost close friends and long-serving comrades. Yet there remained a justified sense of pride in what they had accomplished. In his post-war account, Colonel Wylly reflected on the devastating impact these casualties had on the officers and men of the 9th Yorkshires:

> 'Many of the bravest and best who had served with the Battalion since its formation had been lost forever. When the Battalion fell in at Bellevue [*sic*] Farm on the morning of the 12th July to march back for a well-earned rest, the gaps in the ranks were brought home to the few survivors. It required the utmost determination and self control on the part of all to prevent personal feeling and sentiment ruling the day, and to be able to set out with renewed fearlessness and energy on the great task of reforming a Battalion which should be worthy of those who had fallen, and make it fit to carry on the great traditions they had given and the magnificent example they had set.'[64]

Chapter 13

A Great Loss to the Battalion

Morning broke reluctantly over the rooftops of Harrogate on Sunday, 16 July 1916. Dark and overcast, with ominous banks of low cloud, there was a distinctly autumnal chill in the air and a slow but steady drizzle that lingered well into the evening. It did little to lighten the melancholic mood that had hung over the town since news of the horrific losses on the Somme had first filtered back across the Channel, despite the more encouraging reports of recent days.

The BEF had suffered a staggering 57,470 casualties on the first day of battle alone – almost 20,000 of them killed or died of wounds – and would continue to lose men at an alarming rate. By the end of the month, more than 155,000 soldiers from Britain and her dominions were dead, wounded or missing. For an offensive launched amid such bravado and bellicose jingoism, the scale of loss was nothing short of catastrophic.[1]

Much of the early fighting had involved troops drawn from the industrial towns and cities of northern England and Scotland, and they had paid a heavy price. Yorkshire, in particular, suffered disproportionately on 1 July, with the proud Kitchener battalions of Bradford, Leeds, Barnsley, York and Sheffield sustaining some of the most grievous losses.[2] As a consequence, working-class communities across the county had been plunged into collective and concentrated grief.

The atrocious number of lives squandered, for modest territorial gains, dealt a hammer blow to the British. It also forced Haig and General Sir Henry Rawlinson, the Fourth Army Commander, to reassess their strategic approach. The result was a more carefully coordinated attack, launched at dawn on 14 July and later known as the Battle of Bazentin Ridge, which achieved some tactical success and offered renewed hope for future operations.[3] But the dead still lay thick across the battlefield, while the wounded continued to pour into hospitals around the country in their thousands each day.

At the casualty branch of the War Office, such appalling loss of life presented fresh challenges, for staff were now tasked with the responsibility of notifying vast numbers of families that their loved ones were dead, wounded or missing.[4] Still, by the summer of 1916, the British state possessed a bureaucratic apparatus that had grown exponentially over the course of the war to meet such demands, although the methods of doing so differed according to rank.

Army Form B104-82 brought the devastating news to the families of dead non-commissioned officers and other ranks. It was painfully terse, containing the barest of details.[5] No elaboration, no comfort; just a grim assertion of fact. The next of kin of officers, by contrast, received a War Office telegram, slightly more personal but still starkly dispassionate. Both cut deep and left lasting scars. 'It opened a shapeless horror of greyness,' recalled one bereaved mother, 'in which there was no beginning and no end, no light and no sound.'[6]

Smith and Annie Bell were preparing to leave for Sunday worship on that dark July morning in Harrogate when a boy on a bicycle arrived at Milton Lodge, a telegram clutched tightly in his hand. Don had been a prolific letter-writer, maintaining regular correspondence with his family since embarking for France at the end of 1915. But more than a week had now passed without word, and his mother's unease was deepening by the day. Sadly, her trepidation had not been misplaced as they read the contents of the telegram:

> 'Deeply regret to inform you that 2/Lt D.S. Bell 11 Yorkshire Regt was killed in action 10th July. The Army Council express their sympathy. Please supply name address relationship next of kin.'[7]

The message was devastating in its brevity. Annie Bell, overcome by grief, sat silently on the stairs, staring at the slip of paper. Her husband, characteristically stoic but no less distraught, turned at once to correspondence and his immediate responsibilities. The most pressing, and perhaps most heartrending of all, was the duty of informing his daughter-in-law, Rhoda, that her husband was dead. The couple had been married barely six weeks and were granted precious little time together before Don's leave had ended and he returned to France. Rhoda was at the home of her aunt and uncle in Wilmslow, leaving Smith Bell little choice but to send the dreadful news by telegram. Utterly heartbroken and desperate for more information, she made immediate plans to return to Yorkshire, crossing the Pennines the next day.

On 18 July, Smith Bell responded to the War Office's request for his daughter-in-law's name and address.[8] The following day, Rhoda sat down and wrote her own remarkably composed letter as she began the daunting task of retrieving Don's effects:

> To the Secretary, War Office,
> The news has been conveyed to me by my father in law (Mr Smith Bell) of the death of my husband, Second Lieutenant Donald. S. Bell, on June [*sic*] 10th. 16.
> I enclose my marriage licence and will be glad if you will please forward me his papers etc. as his next of kin, to
>
> Mrs D.S. Bell.
> C/o Mr S Bell
> 87 East Parade
> Harrogate
> Yorks.
> I shall be going to the above address tomorrow July 20th.
> I remain Yours Truly
> Rhoda M. Bell.[9]

The responsibility of ensuring the personal effects of a deceased officer reached his next of kin usually fell to the commanding officer or adjutant of his battalion. Unfortunately, this process was not always straightforward, nor was it necessarily coherent. They first had to be sorted, usually by the officer's assiduous batman, before being packed into his valise and forwarded to the Adjutant General's Office at GHQ in Rouen. From there, they were shipped back to Britain through the French port of Le Havre and would most commonly end up at the depot of Messrs. Cox & Co., Shipping Agency, on Old Burlington Street in Mayfair. They would then be dispatched by rail to various parts of the country before final delivery to the home address of the next of kin.

It was not uncommon for multiple parcels to arrive unannounced on the doorstep of a widow or bereaved parent, with little regard for tact or sensitivity.

Some relatives were horrified to find torn, bloodstained uniforms among the contents; others received shattered watches, photographs shredded by shrapnel or personal letters smeared with dried blood. Such deliveries were deeply traumatic, leaving an indelible mark on those who received them.[10]

Sometimes, personal effects did not reach home at all.[11] Though most soldiers did what they could to recover the possessions of their closest comrades, it was not always possible. Many of the dead received only the most perfunctory of burials, particularly during or immediately after battle. In such circumstances, there was usually little time, and often little inclination, to perform anything more than a cursory search for identity discs or pay books. Even when a thorough search was possible, there was no guarantee that personal items would survive long enough to be officially recorded and returned through the proper channels. Bodies were frequently stripped of badges and buttons by souvenir hunters before burial, while the theft of more valuable possessions like rings, watches and lockets was not unknown, despite constituting a serious breach of military law. Such acts were far less common when burials were conducted by men from the same battalion or regiment, but even then, it was by no means assured that they would not take place. It remains an uncomfortable fact that many families were left distraught when missing possessions could not be satisfactorily accounted for.[12]

After the capture of Contalmaison, Bell's trusted batman, Private John Byers, faced the sad task of packing his belongings into his valise for return to Harrogate. Like many officers, Don had formed a close bond with 'his man', and the pair held one another in high regard:

> 'You wish to know who my servant is as you know someone from Guisborough. They may know him although I doubt it. His name is Byers: W, married about 35, 3 children, lives in 3 West Row, Belmangate, Guisborough, is short & tubby & goes by the name of "Bunny" after the cinema star [American movie actor John Bunny].'[13]

Byers had been at the opposite end of the village when Don was killed and did not learn of his death until the following day. By then, it was too late to recover any of his effects, something he would later admit caused him lasting regret. 'I know that he had his pocket wallet and book with about 30 francs on him at the time,' Byers wrote in a letter to Rhoda. 'His wrist watch, revolver, photo case and several other things that he valued a lot, what became of them I cannot tell.'[14] Whatever suspicions he carried, he kept them close, unwilling to 'cast a slur on some British Regt'.[15]

Among the items Byers did send home were several souvenirs taken from the battlefield at Horseshoe Trench on 5 July. These included a German *pickelhaube* helmet – always highly prized – a bayonet and a pair of boots. Bell's Bausch & Lomb field glasses and leather case were also returned and are now held at the Green Howards Museum in Richmond.

Several months after her husband's death, Rhoda Bell wrote to Byers. His reply, poignant and heartfelt, offers further evidence of the high regard in which Don was held within the battalion:

> 'I would to God that my late master and friend had still been here with us, or better still, been at home with you. I see by your kind and welcome letter that my wife has sent on to you the letter that I sent to her. I think by that you will see what my feelings are to your late husband, my master and platoon officer, because it was not intended for anyone else but her. Believe me, dear madam, what I told my wife was just the truth, if there is any debt, it is on my side and that is a very deep debt of gratitude, also all my company that were left. They worshipped him in their simple, whole hearted way and so they ought, he saved the lot of us from being completely wiped out, by his heroic act. We have lost the best officer and gentleman that ever was with the battalion and we have lost some good ones… The last time we were on the Somme, some of our lads came across Mr Bell's grave and they told me that it was being well cared for, and that there is a cross erected over it, but that is little recompense for the loss you and we have sustained, he was irreplaceable.'[16]

News of Bell's death broke in local newspapers on 18 July 1916. The Bell family were well known and highly respected in Harrogate, and the announcement was met with widespread sorrow. Nowhere was the loss more deeply felt than at the Wesley Chapel, where many had known him since he was a child. In the weeks that followed, heartfelt tributes were offered during services and private gatherings, reflecting the genuine and shared sense of grief.

Harry Angus, who would marry Dorothy Bell in 1921, was devastated when he heard the news and immediately sent his condolences. 'I cannot hope to express all that I feel but I'm sure you all know what Don was to all of us,' he wrote in a letter from Somerleyton in Suffolk. 'I cannot express all that I would say – but Don was always looked up to by those who knew him and we shall all miss him.'[17]

There was also a profound sense of loss at Starbeck School, where Bell had become a much-loved and admired member of staff since taking up his post as schoolmaster in 1911. His death left a lasting impression on all connected with the school, particularly the children. Decades later, one former pupil recalled: 'He was a very good teacher, stern but very popular. He was a fine sportsman and a person who commanded respect.'[18] Another remembered: 'There used to be a photo of Donald Bell along with a plaque in the school hall and each of us had to salute it whenever we went past.'[19]

In London, where Bell had spent two formative years at Westminster Training College, news of his death prompted a flurry of emotional tributes, including one particularly stirring eulogy published in the *Westminster War Bulletin*:

> 'He came to the College with a high reputation for sports, and became Captain of the College athletic team, and a member of the first cricket, rugger, soccer, hockey, and swimming teams of the College. With such a record it is easy to understand Donald Bell's great popularity with his fellow students; and his manly character and fine qualities endeared him equally to the staff.'[20]

Sadly, four years of war would exact a heavy toll on the college. Though it lacked an established Officer Training Corps and the long-standing military traditions of older universities and public schools, more than one hundred Westminsterians would eventually lay down their lives for King and Country.[21]

Given Bell's rising stock at Bradford Park Avenue, it is unsurprising that the *Bradford Daily Telegraph* were among the first newspapers to report that he had been killed. He was a popular figure at the club and had shown enough potential to suggest a promising career lay ahead:

> 'Official intimation has been received of the death in action, last Saturday [*sic*], of Second-Lieut. Donald Bell, of Harrogate. Donald Bell was the Bradford full-back, who played regularly in the reserve team, and on several occasions in the League eleven, with considerable success. A finely-built young back,

> much had been expected of him. He enlisted soon after the war broke out, being the first of the Bradford professional players to ask and secure release in order to go and do his duty and subsequently secured a commission. A few days before his death his battalion took a strong German position, and, owing to the loss of officers, Donald Bell was for a time left in command.'[22]

Tom Maley was deeply saddened by Bell's death and wrote an eloquent tribute to his old full back. 'He played many a fine game for both our teams,' Maley recalled. 'At Nottingham against Notts County he played grandly, but the best of his games was that against the "Wolves", where he completely eclipsed Brooks and Co.'[23] He went on:

> 'In a letter to me he said: "I should think the silly and inane outcry against football is now quashed. I am very pleased to see that most people have realised that its continuance and pursuit is beneficial. You would smile to see us hunting for a field on which to play. No permission is asked. I am afraid if you saw me ready stripped for a game you would say, Donald, laddie, you want rendering down." Poor chap! His hour has come. He has triumphed, and if blameless life and unselfish and willing sacrifice have the virtue attached with which they are credited, Donald is in the possession of eternal happiness, and in his glorious record and great reward there is much to be envied. He was a fine fellow and a good friend, and whilst I regret his loss I feel privileged in being able to place him amongst my boys; my big band of heroic and noble boys. God bless them, every one.'[24]

On 19 July, Billy Bell received a telegram informing him that his brother had been killed in action. While he may have already heard the terrible news, official confirmation left Billy so distraught that he was unable to carry out his duties that day: 'Telegram about Don's death,' he wrote in his diary that night. 'Stay on lorry all day. Too upset to work.'[25]

Several days later, Smith Bell received a letter from Lieutenant Colonel Holmes. Written at Belle Vue Farm on 15 July, it outlined the role Don had played in the fighting at both Horseshoe Trench and Contalmaison. Holmes was unequivocal in his belief that the battalion's success had only been made possible by the gallant actions of his son, deeds he considered deserving of the highest military recognition:

Dear Sir,
I very much regret to have to write and tell you that your son, 2nd Lieut. D.S. Bell, was killed on July 10th. He is a very great loss to the battalion, and also to me personally, and I consider him one of the finest officers I have ever seen.

On the 5th July the battalion made an attack on the German trenches, and almost immediately after leaving our own trenches came under a heavy fire from a concealed machine-gun on our left flank. Seeing that this gun would hold up the attack, your son crept forward with two men and put the gun out of action with a bomb, which also knocked out the gun team. This was a most gallant act, and in my opinion it was entirely due to it that our attack was completely successful. The Germans would not face our men's charge. Many were killed and over 140 prisoners taken, also two machine-guns, and the position was carried.

On the 10th this battalion, with another of the Yorkshire Regiment, attacked and captured a very important village. The Germans again did not stand, and large numbers of prisoners were taken and eight machine-guns, etc. Unfortunately our numbers who attacked the village were few, and the Germans endeavoured to come back and work in behind us. Your son headed a bombing party which drove off the German attack, and once again saved the situation. Unfortunately he was himself shot. Anything more gallant then these two acts cannot be imagined, and he undoubtedly would have received high reward.

I can assure you that you have the sympathy of every officer and man in the battalion. I thought it best to write to you first and not to Mrs Bell, your son's wife, but I will, of course, write to her if she would care to hear from me.

Yours truly,
H.G. Holmes, Lt.-Col.[26]

Still reeling from the death of her husband, Rhoda Bell made an emotional visit to Starbeck School at the end of the month. There, she met several of Don's former colleagues and spent time with children from his class, reading them one of the many letters he had sent to her from France. A photographer from the *Daily Sketch* accompanied her on her visit, capturing events for a full-page spread published on 15 September 1916.[27] Several weeks later, it was announced that a memorial tablet would be erected at the school in honour of their former teacher.[28]

Over the coming weeks, several letters appeared in the *Harrogate Herald* from local men serving at the front, many of whom had known Don before the war. All expressed their sadness and offered their condolences to Rhoda and the Bell family. 'I have just been reading about my old school pal, Don Bell,' wrote W. Outhwaite from Ward 27A of Killingbeck Military Hospital. 'Mrs Bell has my deepest sympathy in her great loss. We were both members of the Y.M.C.A. and went to St Peter's School.'[29] Another Harrogate soldier, W. Knowles, wrote: 'I was very sorry to read the sad news of Don Bell, and I am quite sure it must have come as a great shock to all people who knew him. He was without doubt one of the best.' Private W. P. Birkinshaw, an old school friend serving in France with the Royal Army Medical Corps, added: 'I am trying to find his grave, so that I can put flowers on it. I have seen most of the men in his company, and all of them are very upset about him.'[30]

Of all the letters, the one that perhaps best typifies the response of those who had known Don was sent by Private P.J. Cullingworth and appeared in the paper on 2 August:

> 'I was so sorry to hear of the death in action of 2nd Lieut. Donald Bell. He used to be [a] very intimate friend of mine when he was at Starbeck School. I was living in hopes of meeting him out here in France, but Fate has decreed otherwise. He has done his duty as a soldier and a gentleman. Given his all for his country, so that Britons may live. Harrogate has lost one of its best athletes, for as a football player he was second to none, as the saying goes. I don't know his wife or his relatives, but they have my deepest sympathy in their great loss.'[31]

On 15 August, Rhoda finally received her husband's valise from France. Inside, she found some of his personal belongings, which no doubt brought some degree of comfort, but his private papers were notably absent. This compelled her to write back to the War Office, as she could not settle any of his financial affairs without them.

In France, Billy Bell had been making his own inquiries into his brother's death. At the beginning of August, he had written to Lieutenant Colonel Holmes seeking more information. Holmes was no longer in command of the 9th Yorkshires, having recently been replaced by Major Prior. Nevertheless, he was more than happy to provide details that he hoped would assist the family. Billy wrote to Rhoda soon after, enclosing a copy of the letter he had received from Holmes. It read:

> 'In reply to your letter of the 2nd August, re your brother, 2nd Lieut. Donald Simpson Bell, who was killed on July 10th, he is buried in Contalmaison, near Albert, on the right-hand side just as one enters the village. I can see no harm in giving you this information now, as we are far away from there. He was buried close to where he fell, and there is a redoubt close there called "Bell's Redoubt" after him. He was a noble soldier and a braver fellow never stepped, and he is an irreparable loss to the battalion.'[32]

The grave was among a small cluster scattered across a field at the south-eastern edge of the village, around 70 yards from the junction of the *Rue de la Carrière* and the *Rue de Mametz*. All were diligently cared for, but Don's grave was particularly well maintained. A white wooden cross, inscribed with his details and the cap badge of the Green Howards, had since been erected, while a wooden picket fence had been constructed around some freshly laid turf by a soldier from the Royal Engineers. His shell-ravaged steel helmet, meanwhile, sat on top of the grave. One soldier noted that it was 'the nicest and cleanest spot about here'.[33]

Back in Britain, Rhoda had received no further communication from the War Office regarding her husband's private papers. She had, however, been notified that her application for a widow's pension had been approved by the Pension Issue Office – news that must have eased her anxieties somewhat. The claim seemingly progressed with minimal delay and was granted on 11 August at a rate of £100 per annum, backdated to 11 July 1916.[34] It would not heal the grief caused by Don's death, but it would alleviate some of the financial strain.

At the end of the month, Rhoda finally received a response to the letter she had sent to the War Office on 15 August. Although it made no reference to her appeal for his private papers, or indeed any money he may have left, it did arrive accompanied by the death certificate she had requested.[35] That was, at least, some small comfort.

Chapter 14

The Most Distinguished and Most Coveted of All

A wave of civic pride swept through Harrogate when news broke that Second Lieutenant Donald Simpson Bell had been awarded a posthumous Victoria Cross for his actions at Horseshoe Trench. It was the town's second such honour in the space of six months, and Harrogatonians were justifiably proud of both recipients. The citation was published in *The London Gazette* on Friday, 8 September 1916 – eight weeks after his death:

> 'For most conspicuous bravery. During an attack a very heavy enfilade fire was opened on the attacking company by a hostile machine gun. 2nd Lt. Bell immediately, and on his own initiative, crept up a communication trench and then, followed by Corpl. Colwill and Pte. Batey, rushed across the open under very heavy fire and attacked the machine gun, shooting the firer with his revolver, and destroying gun and personnel with bombs. This very brave act saved many lives and ensured the success of the attack. Five days later this gallant officer lost his life performing a very similar act of bravery.'[1]

It was one of twenty Victoria Crosses gazetted that day, including several awarded to men killed on 1 July. Among the most noteworthy of these recipients were Major Stewart Loudoun-Shand, a company commander with the 10th Yorkshires, who refused to leave his men despite being mortally wounded during their ill-fated assault at Fricourt, and Private William Henry Short, a gifted amateur footballer from Eston, near Middlesbrough. Short was a battalion bomber with the 8th Yorkshires and may well have been present when Bell lost his life on 10 July. The 31-year-old was recognized for his 'great gallantry' at Munster Alley on 6 August when, seriously wounded in the foot, he continued to hurl bombs at the enemy. 'Later his leg was shattered by a shell, and he was unable to stand,' reads his citation, 'so he lay in the trench adjusting detonators and straightening the pins of bombs for his comrades.'[2] Sadly, Short died before he could be carried out of the trench and was later buried in the grounds of the ruined château at Contalmaison.

Among the recipients who survived was the indomitable Lieutenant Colonel Adrian Carton de Wiart, the one-handed, eye patch-wearing commanding officer of the 8th Gloucesters who was wounded at least eleven times during a long and distinguished military career that began in the ranks of the Imperial Yeomanry in 1900. Dubbed the 'unkillable soldier', Carton de Wiart was recognized for 'most conspicuous bravery, coolness and determination during severe operations of a prolonged nature' at La Boisselle between 2 and 3 July.[3]

One week after the announcement of Bell's award, the *Harrogate Herald* published a full obituary under the headline 'Another Harrogate VC':

> 'Harrogate people deeply regret that Second Lieutenant Donald S. Bell does not live to wear the Victoria Cross, which his gallant conduct secured him a few days before his lamented death…
>
> 'He was well-known in Harrogate and a school teacher under the Harrogate Corporation when war broke out. Since his death the Corporation have passed a resolution putting on record their high appreciation of his services as a school teacher and also as a soldier of the King. He enlisted as a Private in the West Yorkshires when the war broke out, and became a non-commissioned officer, and later was given a commission in the Green Howards. In the early days of the "Big Push" he was second-in-command of a section, and when his senior was knocked out he took command. He spotted a machine-gun on the left which was enfilading the whole of the front, and crawled up a communication trench, accompanied by Private Batey, who has since received the DCM. Lieutenant Bell with his first bomb at twenty yards range hit the machine-gun and put it out of action. He followed this up by knocking out the gun team. He was thanked on the field by his Commanding Officer, and since then a Bell's Redoubt has been named after him.
>
> 'As we intimated at the time Lieutenant Bell performed his gallant and daring action he took a very modest view of the deed, and although it was rumoured at the time that he had been recommended for the VC, we are not allowed to mention these awards until they are officially announced. Lieutenant Bell was a member of the YMCA for many years and one of the most popular members. Three winters ago he was a member of the

> International YMCA Football Team that had a tour of Denmark, when his play won golden opinions. He was also a member of the Claro Tent of the Independent Order of Rechabites...
>
> 'Lieutenant Bell was a keen sportsman, and well-known locally at one time as a flat race sprinter. He was a playing member of the Bradford Park Avenue team when war broke out, and previously played for Newcastle United, Bishop Auckland, Mirfield United, Starbeck, etc. As the majority of readers are aware, Lieutenant Bell was the younger son of Mr and Mrs Smith Bell, of East Parade, Harrogate, and was married to Miss R M Bonson at Kirkby Stephen, Westmorland, only some five weeks before he was killed. This is the second VC which has come to Harrogate during the war, the other recipient being Shoeing-Smith Hull, son of Mr and Mrs Hull, Albert Terrace, Harrogate.'[4]

It was a remarkable coincidence that both of Harrogate's VC recipients had attended St Peter's School. Unlike Bell, however, Charles Hull survived the war and returned home – although he had to wait until 3 December 1919 to receive his medal from King George V.

Across the Channel, the 9th Yorkshires were back in familiar surroundings after a spell on the Franco–Belgian border. The battalion was now under the command of Major Prior, and he wrote to Rhoda Bell from the village of Hénencourt on 13 September to congratulate her on her husband's award:[5]

> Dear Mrs Bell,
>
> On behalf of the officers, non-commissioned officers, and men of this battalion, I write to offer you our most heartfelt congratulations on the honour which your gallant husband has won both for those near and dear to him, and also for this battalion. We are proud to have had him amongst us. During that short time he earned the love and respect of all who knew him, and we join with you in deep mourning his loss. The action which earned him the V.C. was one of noble, self-denying, unflinching courage, and by that noble deed he saved many lives and assured the success of the attack.
>
> Trusting that this highest honour which has been conferred by His Majesty on your late husband may do something to lighten the burden of your grief.
>
> I am, dear Mrs Bell,
>
> Yours very sincerely, H.A.S. Prior, Major,
>
> Commanding 9th Batt. Yorkshire Regiment, B.E.F.

The letter was no doubt appreciated, but Rhoda was still endeavouring to recover Don's personal papers in order to settle his affairs. After a perfunctory response to her last enquiry, she wrote again to the War Office on 14 September, only to receive another frustrating reply. 'I am directed to acquaint you that no effects of the late Second Lieutenant D.S. Bell, 9th Battalion, Yorkshire Regiment, other than those already forwarded to you have been placed at the disposal of this department,' she was informed at the end of the month. 'I am to add that this department has no knowledge at present of any Will having been executed by the late officer.'[6]

Talk of Bell's Victoria Cross dominated morning worship at the Wesley Chapel on Sunday, 17 September. A significant number of its young men were serving overseas, including many still engaged in the protracted fighting on the Somme. Their service was a source of considerable pride at home, though this was tempered by the sombre recognition that not all would return. The chapel contributed directly to the war effort too, opening a soldiers' club in its basement, staffed by volunteers. The club proved immensely popular and was frequented by men stationed at nearby camps. In the latter part of the war, the army also requisitioned the Sunday school premises, which were later used during the period of demobilization following the Armistice.[7]

Thoughts of an end to the fighting were far from the minds of most during that day's service. Don had attended the chapel since childhood, and he was the focus of much of the sermon delivered by the Reverend R.P. Lowe:

> 'The newspapers of last Monday told of certain military honours to be bestowed by the King. One of these – the most distinguished and most coveted of all, the V.C. – comes to a man whose memory this Church will be proud to revere for long years to come. In the annals of Wesley, Lieutenant Donald Bell will live as a strong, wholesome, chivalrous saint. He was the morning superintendent of our Sunday school, one of our Junior Society class leaders, an active member of the choir – in the van of all aggressive endeavour for the widening of the frontier of the Kingdom – as valiant a soldier of the Cross of Saint George. It will be in all our hearts to congratulate Mrs Donald Bell and Mr and Mrs Smith Bell on the honour which becomes theirs – and with our congratulation there goes the uttermost of affectionate sympathy. We do well to remember, as Lieutenant Gladstone said, in a letter written home to his mother, not long before he fell in action – "It is not the length of existence

> that counts, but what is achieved during the existence, however, short" – and judged by that standard Donald Bell has lived long and died grandly. In a letter written to his mother a couple of days after the deed which won this distinction, he said: "Remember, don't worry about me. I believe that God is watching over me, and it rests with Him whether I pull through or not." Of him it may be said – "Surging with life he went, not as a spirit agent, Lost to our sight – be in his joyous might, Leapt to the infinite, Climbed the white steep that yearns Out of our night, Past the world's woe, and burns Bright with his seraph Sight, Brother of mine, God's right.'"[8]

Nine days after Bell's Victoria Cross was announced in the *London Gazette*, the same publication carried the news that Lance Corporal H. Colwill and Private J. Batey were to both receive the Distinguished Conduct Medal in recognition of their bravery that day. Harrison Colwill was presented with his DCM at Elemore Union Hall in Houghton-le-Spring, County Durham, on 9 December 1916. He also received an inscribed gold watch, funded by public subscription, and a book of war savings certificates from the Hetton Colliery Hero Fund.[9] The 26-year-old was a miner before the war but chose to leave his reserved occupation in November 1914 to enlist. He had served with the 9th Yorkshires since their formation and was only appointed lance corporal in the days leading up to the start of the British offensive on the Somme. Colwill came through the attack at Horseshoe Trench unscathed but was wounded in the left leg during the fighting at Contalmaison. He eventually found himself in hospital in Bradford, where he spent a month recovering from his injuries. After leaving hospital, he was posted to the 3rd (Reserve) Battalion at West Hartlepool, where he spent the rest of the war as a physical training instructor.[10] Colwill returned to work at Elemore Colliery following his discharge from the army and died in Sunderland, aged 79, in 1967.

Joseph Batey was a miner too, standing just 5ft 3in tall and weighing 8½ stone when he joined up at the age of 19 in November 1914. His diminutive size belied a fiery character, and his tenacity on 5 July prompted Bell to claim the young private 'did all the work'.[11] Unfortunately, Batey was also something of a hothead and appeared before a court martial on more than one occasion during his four years of military service. He had to wait until 31 May 1918 to receive his DCM, although it was presented to him by George V during a royal visit to Beckett Park Hospital in Leeds. Alas, that did not mark the end of his run-ins with the authorities. Just weeks after being decorated, the disillusioned youngster absented himself from duty and was hauled back in front of another court martial – this time charged with desertion. The War Office tried to take back Batey's DCM following his discharge in 1919. However, despite their best efforts, they were unable to locate the medal; it was presumably kept hidden away until his death in Sunderland in 1969.[12]

On 23 September 1916, Bradford Park Avenue played host to Birmingham in a Midlands Section league game. The club used the occasion to honour Bell and announced all spectators would receive a reproduction signed commemorative portrait of the full back when they entered the ground.[13] Around 3,000 eventually passed through the turnstiles, being treated to a thoroughly entertaining contest.

The home side dominated the opening stages and raced into a two-goal lead. Despite also striking the woodwork twice, Bradford failed to extend their advantage, and the visitors staged a remarkable second half comeback to secure a narrow 3–2 victory.[14] All involved later agreed it had been a match worthy of the occasion, and the day would live long in the memory of those in attendance.

Stirring tributes continued to flood in over the coming weeks, as those who had known Don expressed their respect and admiration for his gallant deeds. Many came from former classmates and childhood friends. 'Harrogate has need to be proud of him,' wrote S. Macklin from HMAS *Australia*, 'for it is such men as he that are keeping the old flag flying.'[15] W. Outhwaite, a former pupil of St Peter's, added: 'You have no idea how proud I feel of my old school chums, as both the V.C. winners were my old school pals, and I am sure our old schoolmaster, Mr Andrews, must feel proud of them too.'[16]

Among the tributes received from Westminster Training College was a letter published in the October issue of the *Westminster War Bulletin*. Written by Sergeant J.R. Crossland, then serving in France with the Royal Garrison Artillery, it reflected the high regard in which Bell was still held at the college:

> 'I have pinned up the portrait of "Don" Bell on my wall here – somewhere in the great "Push" – proud to belong to the college that helped to train such heroes. How well I remember walking round the "Smokes", in the days when we were training for the inter-coll. sports, and looking at "Don" Bell's photo, prominent among the team that won the shield. His prowess was an incentive to us who came after, and it was by emulating his thoroughness that we gained the trophy in 1913.'[17]

On 5 December, Rhoda received a telegram from the War Office requesting her presence at Buckingham Palace. A formal invitation arrived soon after:

> 'Mrs Donald S. Bell, widow of the Harrogate V.C., Second-Lieut. Donald Simpson Bell, has been summoned to appear at Buckingham Palace on Wednesday to receive the decoration won by her late husband.'

It was accompanied by a letter from the king:

> 'It is a matter of sincere regret to me that the death of Second-Lieut. Donald Simpson Bell deprived me of the pride of personally conferring upon him the Victoria Cross, the greatest of all rewards for valour and devotion to duty.
> 'GEORGE R.I.'[18]

On the morning of Wednesday, 13 December, Rhoda travelled by train to London to receive her husband's Victoria Cross from King George V at an investiture ceremony held at Buckingham Palace. She was accompanied by her sister-in-law, Minnie. After the presentation, the two women were photographed at the palace gates, Rhoda holding Don's medal in her left hand. Both wore black hats, gloves and scarves against the chill of a bleak winter's day.[19]

Among the other guests honoured, were two officers from the Royal Flying Corps: Captain William Price, a pilot with No. 11 Squadron, and his American observer, Lieutenant Frederick Libby. Both men received the Military Cross for operations conducted over an eight-week period in France, during which they brought down six enemy aircraft in their F.E.2.b two-seater.[20] The king also presented the insignia of a Knight Commander of the Order of the Bath to Lady Arbuthnot, wife of Rear Admiral Sir

Robert Keith Arbuthnot, and to the Honourable Lady Hood, wife of Rear Admiral the Honourable Horace Hood – both of whom had been killed at the Battle of Jutland on 31 May 1916.[21]

Rhoda appears to have maintained regular contact with her brother-in-law, Billy, for several months at least following Don's death, as both sought to navigate their shared grief. On Christmas Day, he received a letter detailing her trip to Buckingham Palace with his sister. Writing to his mother later that day, Billy commented: 'Minnie and Rhoda seem to have had a good time in London.'[22] In a subsequent letter, he revealed she had also sent him a photograph: 'I had a letter from Rhoda today with a photo of herself and Don.' It was clearly appreciated, as he remarked it was the only 'proper one' of his brother he had. Another letter home suggests Billy was compiling quite the family album: 'Thanks very much for the photos of you, Dad and Dolly. They are very good. There are only Gertie and Minnie who have not sent me one to make up a complete set.'[23]

Billy was billeted at Louvencourt, little more than ten miles from Contalmaison, but he had to wait until 5 January 1917 to visit his brother's grave for the first time. Riding south on a borrowed motorcycle, it took him several hours to locate, but when he did, he was relieved to find the grave meticulously cared for. Before leaving, he removed Don's steel helmet so it could be sent home. In his diary, Billy noted that it had been a 'nice day', though his mood was soured somewhat by a puncture on the journey back to Louvencourt.[24] Nine days later, he wrote to his mother to tell her about the trip:

> 'I think I forgot to tell you that I visited Don's grave before we left the other place which you knew. I went on a motorbike & passed two villages (at least the spots where villages once were). You cannot find any trace of houses as the ground has been churned up over & over. All I could go by when I got to Contalmaison was the fact that the crossroads should be in centre of village. I arrived mid-day & after a 4 hour search found it. It is the nicest grave in the district & there are a lot around. I could not bear to leave his steel helmet so have kept it. When I stood on the spot I could just picture how the fighting went & how easy for Germans to get very near without being seen.'[25]

By coincidence, the *Harrogate Herald* published a letter several days later that reiterated just how well-kept Don's grave was. It was written by one of his old football pals, Sapper Thomas Enderby, who was billeted at Contalmaison with the Royal Engineers:

> 'It may interest you to know that I am living in a dugout just about fifty yards from the spot where Sec.-Lieut. Don Bell, V.C., was killed. He is buried there, and his grave is the brightest spot in the vicinity. Over it is erected a cross, and around it are neat railings. Within these is smoothly laid turf, which is even now quite green and fresh. His helmet rests on the grave… I went to see if I could do anything to it, but it was very nice. It is the nicest and cleanest spot about here. Don Bell was a great friend of mine when I used to play football. I have played with him and seen him play numbers of times. Poor old Don! Everybody liked him.'[26]

In February 1917, a 'very large, framed photograph' of Bell was donated to the town by Mr Thomas A. Cornall, proprietor of the Station Studio, who intended it 'to be hung in the Public Library'.[27] The portrait, which was reported to have been readily accepted, was just one of several gestures by which the people of Harrogate sought to honour their fallen hero.

On the afternoon of Friday, 30 March 1917, Rhoda joined Smith and Annie Bell, as well as other members of the Bell family and a group of dignitaries, for the unveiling of a memorial tablet in Don's memory at Starbeck School. The tablet, shrouded in black and surmounted by a Union Flag, was unveiled by the Mayoress of Harrogate during what was described as a 'sad, yet inspiring ceremony'.[28] It read:

> 'To the memory of Second-Lieut. Donald Simpson Bell, V.C., of the Yorkshire Regiment (Green Howards), a teacher on the staff of the school, who gained the Victoria Cross for a deed of gallantry during the British advance on the Somme on the 5th July, 1916, and who was killed at Contalmaison whilst performing a similar act of bravery on the 10th July, 1916. Aged 25 years.'[29]

After those present had 'impressively sung the National Anthem', Smith Bell stood to thank the Mayor and Mayoress, as well as the Harrogate Education Committee, for erecting the tablet. He added that there were 'thousands like his son at the Front ready to lay down their lives for King and Country'.[30] The Mayor of Harrogate concluded proceedings by 'expressing the hope that the tablet would be an incentive to the boys and girls in their games and work'.[31]

Further tributes would follow in the ensuing months, including a proposal to establish a 'V.C. Scholarship' to support the education of promising pupils at Bell and Archie White's former school on Haywra Crescent. 'The appropriateness of the title may be judged from the fact that two students at the Secondary School have gained V.C.'s in the present war,' reported the *Leeds Mercury*, adding it was hoped the initiative would help 'perpetuate the names of those who have earned the coveted honour'.[32]

There had been significant developments in France in the weeks leading up to the unveiling of the memorial at Starbeck School. On the Somme, the German Army had fallen back, though not in defeat. Operation Alberich was a strategic withdrawal to fortified positions along the shorter, thus more easily defended, Hindenburg Line.[33] Having fought so bitterly for every inch of ground in 1916, many British soldiers reacted with resentment to the withdrawal, believing it rendered the appalling loss of life all but meaningless.

The advance towards the Hindenburg Line posed substantial logistical challenges too, with Billy Bell and the ASC playing a critical role.[34] 'I have been on the go since yesterday morning carrying ammo all night,' he wrote to his parents. 'We look like being busier than ever during the next few days.' He had a narrow escape on one journey, revealing that a 'bit of shell' had struck his lorry but thankfully caused no damage, other than breaking two boxes inside. 'My luck still keeps good,' he quipped.[35]

At the end of June 1917, another portrait of Bell was unveiled in his hometown, this time at the Wesley Sunday School to mark the centenary of its founding:

> 'Last night advantage was taken of the Sunday school anniversary to pay further tribute to the memory and gallantry of Second Lieutenant Donald S Bell, VC, at Wesley Schoolroom, where a portrait of the lamented officer was unveiled by the Rev JR Irving, a former resident minister, who, along with the Rev RP Lowe, and Lieutenant FT Kettlewell bore testimony to the late soldier's sterling worth and character. The portrait, by Mr TA Cornall,

Station Square Studios, Harrogate, is an excellent likeness of the deceased officer and is enclosed in a dark oak frame. Beneath is a brass tablet with the following inscription: "A good soldier of Jesus Christ. To the memory of Second Lieutenant Donald S Bell, VC, who fell in action at Contalmaison July 10th, 1916. A valued teacher of the Wesley Sunday School. Greatly loved by all." This has been neatly engraved in old English and plain letters with red initials and black border by Mr FB Jesper, Prospect Crescent, Harrogate.'[36]

Billy Bell returned to Contalmaison on 19 July 1917, taking three lorries and a party of men from his company to assist the Royal Engineers with salvage work. While there, he spent several hours tending to his brother's grave and noted in his diary that the Graves Registration Unit had recently affixed an aluminium identity plate to the cross.[37] He described the scarred landscape as like 'being up at Pateley Moors', though he added a grim caveat: 'Only 30 yards away a German's legs are showing out of the earth.'[38] Some months later, Billy sent some photographs of the grave home in a letter:

'I am enclosing some more snaps of Don's grave which another fellow did for me. They give you a better idea of the surroundings. The peculiar thing is, that that is the only bit of hedge you can find for miles for there are no hedges or walls in that part of the country. You cannot make out the shell holes for the long grass makes the ground look level. Please send me an enlargement if finished for after I've seen it I wish to present it to this fellow who has often obliged me in other ways.'[39]

By the turn of the year, Billy had resolved to put himself forward for officer selection. He passed a medical examination with ease at the end of January and was soon being interviewed by a brigadier general at headquarters. 'I am waiting for the result,' he told his parents in a letter dated 3 February 1918, adding:

'If he is satisfied with me I shall probably go to a base workshops & have 3 or 4 weeks test in a shop & sit for an examination. Then I shall interview a General & if I pass him, will be sent to London to finish off the cadets' training. If I get so far I shall then be put on a 6 months' probation as 2nd Lieutenant. There is a lot of messing about for if I finished my training the General could reject me at the finish. Still I am hoping for the best & and swotting away at theory… I am due to leave here on 19th of this month for my leave but if I am successful with commission stunt shall probably be away from company by then & lose it. But in any case I ought to be home before very long.'[40]

Two weeks later, after months of speculation and no shortage of grumbling, Billy finally returned home on leave again, though it was all too brief. By 5 March 1918, he was back with his company, but not for long, as a letter written on St Patrick's Day reveals:

'Here I am in the middle of my preliminary test down at G.H.Q. Supply Column. I left my unit 5.45am Friday morning. Had to change 3 times & stay overnight at the last place. Of course I found very good billets in company with a sergeant. Had a good tea at the YMCA. Had a better supper at the billet & bed each with a jolly good breakfast of ham & eggs the following day. Not at all dear either. Arrived 11am Sat morning after travelling down with French civilians. I was able to practice my French & quite enjoyed the whole journey. As soon as we arrive, still more interviews with same old questions. Sat afternoon we go in workshops & have a job given us. Our hours 7.15am parade, Work from 8.30 to 12.30 & 1.30 to 5.30 with evening parade 9pm. This morning we sat for a 3 hours written examination at which I was not at a loss. I understand we may go before the General on Wed. when I shall get my answer or otherwise. It is all a matter of luck as I have explained to you before.'[41]

Unfortunately, Billy's efforts to become a 'temporary gentleman' would be thwarted. 'I am once more a failure,' he lamented in a letter home on 1 April, though he did not elaborate on the reason why.[42]

There were, however, far greater concerns. In the early hours of 21 March 1918, the Germans launched their long-expected spring offensive, also known as *Kaiserschlacht* ('Kaiser's Battle'), in a final gamble to break the deadlock before the newly arrived American Expeditionary Force was fully trained and equipped for combat. Made possible by the transfer of troops from the east after Russia's defeat and collapse into revolution, the first phase, Operation Michael, struck from the Hindenburg Line with the objective of driving a wedge between the British and French armies near the River Somme.

Preceded by a brutal five-hour artillery bombardment, the offensive achieved dramatic initial success. Allied troops were ordered to retreat, and thousands surrendered. Within days, Bapaume and Péronne were lost. Then, for the first time in four years of war, the town of Albert fell as German troops swept across the old Somme battlefield – including Contalmaison, which fell without a fight.[43] By 5 April, the enemy had captured 1,200 square miles of ground, advanced to within eleven miles of Amiens and taken more than 75,000 British prisoners.[44]

Operation Michael marked the opening phase of the German Spring Offensive, with subsequent operations launched in Flanders, on the Aisne and along the Marne. Yet despite achieving significant territorial gains and driving deep into Allied lines, the offensive failed to deliver a decisive strategic breakthrough. German losses were also heavy, many of them irreplaceable elite *Stoßtruppen* ('Stormtrooper') units, and their supply lines overextended. This left them occupying forward positions that were both dangerously exposed and increasingly vulnerable to attack.[45]

On 8 August 1918, the Allies struck back. In coordination with French forces, British and Dominion troops launched a major offensive astride the River Somme, east of Amiens. By the end of the first day, a gap fifteen miles wide and up to ten miles deep had been punched through the German line south of the river, spreading panic and confusion.[46] General Ludendorff later called it 'the Black Day of the German Army',[47] as dramatic Allied gains triggered a collapse in the morale of his troops, who surrendered in their thousands. 'As the sun set over the battlefield on 8 August,' notes the official account, 'the greatest defeat sustained by the German army since the start of the war was an accomplished fact.'[48]

The Battle of Amiens, later recognized as the beginning of the Hundred Days Offensive, marked the start of a relentless Allied advance that steadily drove the Germans eastward and eventually brought about the end of the war. But progress came at a price. More than 410,000 soldiers from Britain and its Empire were lost between July and November, as the enemy conducted a series of disciplined and often ferocious rearguard actions.[49] In fact, the BEF's daily rate of loss during this period was significantly higher than it had been during both the Battle of the Somme in 1916 and Third Ypres a year later.[50] Among the dead during these final few months of the war was

Don and Billy Bell's cousin, George Simpson, who fell near the village of Foucaucourt on 23 August 1918.

In October, news reached Yorkshire that one of Bradford Park Avenue's best players, Jimmy Smith, had also been killed.[51] The death of the popular centre forward affected everyone at the club, but none more so than his manager, with whom he had remained in regular contact. 'He told me of his intention to get married,' Tom Maley said in a heartfelt tribute, 'and how he longed for the dawn of the day which would see him back again at the Avenue.'[52]

Among the other notable deaths that month was Lieutenant Colonel Bernard William Vann, the commanding officer of the 1/6th Sherwood Foresters. The 31-year-old, an ordained clergyman, was an accomplished amateur footballer who had made several appearances for Derby County and Burton United before the war. Vann was killed four days after leading his men across the Canal du Nord in an action for which he was posthumously awarded the Victoria Cross, thus becoming one of only two VC recipients – alongside Donald Bell – to have played in the Football League.[53]

Little more than four weeks later, the war was over, and it ended as abruptly as it had begun. At 11.00 am on 11 November 1918, the guns stopped firing, and a mantle of 'unwonted silence'[54] descended over no-man's land. Jubilant crowds poured onto the streets at home, mounting impromptu celebrations, while church bells rang out across the land. Yet for most at the front, the moment was marked not by elation but by relief and sheer exhaustion. Too many friends and comrades had been lost over four grinding years of conflict, and uncertainty loomed over what lay ahead. 'It is all over but we have no time for joy making,' Billy wrote to his mother the following day. 'I am guessing the date of my discharge. I say about February with luck… I am wondering what I shall do after peace?'[55]

On 10 January 1919, Billy Bell left France – and the grave of his younger brother – for good, and sailed home across the English Channel. He would spend the next month at North Camp in Ripon before finally being demobilized and transferred to Army Reserve Class 'Z' on 8 February.[56] After almost four-and-a-half years in khaki, Billy returned to civilian life in Harrogate, his war over at last.

Chapter 15

Epilogue: Memory, Loss and Legacy

In July 1919, men from the 21st Labour Company exhumed Donald Bell's bodily remains and transported them one mile west for reburial at Gordon Dump Cemetery. It was one of more than 200,000 exhumations carried out by units of the British Army, under the command of the Directorate of Graves Registrations and Enquiries, between the Armistice and the end of 1921.[1] It was a grim but necessary task, for the remains of countless thousands of soldiers, both identified and unknown, were scattered across vast tracts of the battlefield; some in small ad hoc cemeteries, others in lonely graves, many not marked at all.

The majority of those exhumed were moved to newly created 'concentration' cemeteries or existing sites that had been significantly expanded by the Imperial War Graves Commission. Gordon Dump Cemetery was first used after the fighting of 10 July 1916 but closed two months later, with a total of ninety-five burials. It was reopened at the end of the war when bodies were brought in from the surrounding area. Today, 1,676 dead are buried or commemorated there, though just 624 of these are identified, which gives some indication of the devastation wrought in the area.[2] Aside from special memorials to thirty-four casualties known or believed to be buried among them, the remainder lie under the epitaph 'Known Unto God'[3] – chosen by lauded writer, and mythologizer of Empire, Rudyard Kipling, who lost his only son at Loos in 1915.

Gordon Dump is a quintessential British war cemetery. Its neatly manicured lawns, stately trees, shrubs and herbaceous borders embody the 'homely sense of the English churchyard' envisioned by its principal architect, Sir Herbert Baker.[4] The Cross of Sacrifice, designed in 1918 by Sir Reginald Blomfield, stands sentinel over the cemetery's regimented rows of Portland headstones, while Sir Edwin Lutyens' Stone of Remembrance lies in solemn reverence at its centre. Its inscription, 'Their Name Liveth for Evermore', was taken from Ecclesiasticus 44:14 at Kipling's behest.[5]

A sepia postcard, produced in the immediate post-war period, reveals the cemetery in its formative state. It dates back to before the original wooden crosses were replaced by permanent headstones, perhaps 1920 or 1921. The Cross of Sacrifice and Stone of Remembrance have yet to be erected, while the barren landscape is devoid of shelters, trees and perimeter wall. Taken from the south, where a track once linked La Boisselle and Contalmaison, the wartime light railway that ran through the valley is visible in the foreground. The cemetery's original entrance was here, though all physical traces have long gone. Today, access is only possible from the north, where the main road provides an excellent vista across Sausage Valley and the surrounding battlefield. There is a poignant symmetry to Donald Bell's reburial here, for just a few hundred yards beyond the cemetery, over the horizon, lies Horseshoe Trench, site of his heroic deeds of 5 July 1916.

As Britain grappled with the profound sense of loss following the Armistice, communities across the country sought tangible ways to commemorate the vast army of the dead. These acts of remembrance were both deeply personal and collective, reflecting the grief, gratitude and solemnity of a nation traumatized by an unprecedented scale of bereavement. In time, nearly every town, village, workplace, church and school would boast its own memorial, ranging from the understated to the grandiose. Most were erected in the decade after the Armistice, but some went up while war was still being waged. Among the very earliest were the 'street shrines' found in proud working-class communities: makeshift memorial boards erected to celebrate the extraordinary response of their menfolk to Kitchener's call to arms. Sadly, they would become memorials to the dead soon enough.[6]

On 19 July 1919, some 20,000 men and women from Britain and its Allies marched through the streets of London as part of the nation's Peace Day celebrations. During the parade, 15,000 soldiers and 1,500 officers proceeded down Whitehall, passing a temporary 'Cenotaph' erected outside the Foreign Office to serve as the focal point of the commemorations.[7] Designed by Lutyens, the memorial had the appearance of stone but was constructed entirely from wood, plaster and canvas. It was also due to be dismantled shortly after the event. However, the Cenotaph quickly became a point of pilgrimage for grieving relatives. Within a week, an estimated 1.2 million people had visited the structure to lay flowers and pay their respects. *The Times* reported that 'no feature of the victory march in London made a deeper impression than the Cenotaph'.[8]

Thousands gathered there again on 11 November 1919 – the first anniversary of the Armistice – when a two-minute silence was observed following a march-past by veterans and a wreath-laying ceremony. The temporary memorial had such a profound impact that the Liberal government resolved to reconstruct it in stone. Work on the permanent structure began in May 1920 and was complete five months later. On 11 November 1920, King George V unveiled the new Portland stone Cenotaph as the United Kingdom's national memorial to the war dead of Britain and the British Empire.[9] The unveiling took place on the same day as the ceremonial repatriation of the Unknown Warrior at Westminster Abbey.[10]

In the summer of 1920, Rhoda Bell received a second invitation to Buckingham Palace – this time to attend a Royal Garden Party held in honour of the nation's Victoria Cross recipients. Smith and Annie Bell accompanied her to the event on 26 June, which marked the largest official gathering of VC holders since Queen Victoria decorated the first sixty-two recipients at Hyde Park on the same date in 1857.

As part of the celebrations, those able to do so assembled at Wellington Barracks and marched along Birdcage Walk and Horse Guards Parade before proceeding down The Mall to the palace, where they were met by George V and Queen Mary. 'Mr Smith Bell was struck by the simplicity and humanity of the party,' reported the *Harrogate Herald* several days later, 'and he speaks in high terms of the King's great interest in the men.' Had his son lived, the paper added, the day 'would have been the happiest in his life'.[11]

Don's old school friends, Archie White and Charles Hull – by then a serving police officer with Leeds Constabulary – were also in attendance. Hull's mother was reported to be 'delighted that His Majesty should remember the circumstances under which her boy won his VC'.[12] Hull died in 1953 at the age of 62 and was buried with full military honours.

In Harrogate, efforts to commemorate the sacrifices made by the town's young men were gathering momentum. At the Wesley Chapel, £250 had been raised for a memorial to honour 'all Wesley boys who had served in the War'.[13]

Erected in the vestibule, the large brass plaque, with its ornate surround, bears the names of eighteen men from the chapel who lost their lives, along with a further 114 who served but came home. Don is listed among the fallen; his brother Billy appears among the returned, alongside Minnie's husband, Tom Wood, and Dolly's future husband, Harold Angus. A donation of £100 from the remaining funds was given to Harrogate Infirmary.[14]

Beneath the main memorial today are two smaller plaques. One, erected in the 1980s, commemorates long-serving organist and choirmaster Harold Attley. The other is much older and was originally located at the Sunday school at the bottom of Cheltenham Crescent. It reads:

> 'A Good Soldier of Jesus Christ. To the Memory of 2nd Lieut. Donald S. Bell V.C. Who fell in action at Contalmaison July 10th 1916. A valued teacher of Wesley Sunday School. Greatly loved by All.'[15]

Across the country, the memorialization of the dead continued. At Westminster Training College, nearly £1,000 was raised through subscriptions and donations for a memorial to honour the 102 Old W's who had died during the war. This effort culminated in the dedication of two large oak memorial tablets in the college chapel on 10 July 1921 – the fifth anniversary of Bell's death on the Somme. The service was attended by a large gathering of past and present students, staff and relatives of the dead, who 'joined together to remember those who had first reached their journey's end'. Don's name is among them, along with those of several of his closest friends and contemporaries.[16]

That 'right champion fellow' Bertram Stanley Temperton, whom Don first met on the train down to London in 1909, went missing during heavy fighting north of Contalmaison on 29 July 1916 and was never seen again. Ralph Farnham, his high jump rival, died of wounds on 31 October 1918, just eleven days before the guns fell silent. Among the others, Archie Robertson, another gifted athlete from the sports team, was wounded at Serre on 1 July 1916 and died three weeks later. Ronald Glisbey Ayliffe, Harrison Milton Bailey, Hubert George Maidment and Richard Hanson Newbold all fell on the Somme in September 1916, while John Edwin Newton was killed on 21 September 1918. Twelve students from the 1908–10 cohort also lost their lives during the war, including Charles Henry Saunders and Fred Holloway, whom Don had competed with and against on the athletics field.[17]

A stained-glass window was also unveiled in the chapel that day in Bell's honour, with the remaining memorial funds vested in a trust to support the education of children whose fathers had been killed in the war.[18] Both the memorial tablets and the window were placed in storage for safekeeping during the Second World War – a prudent decision, as the college sustained significant damage during a Luftwaffe raid on the night of 4 March 1944. The chapel alone suffered almost £5,000 worth of damage, largely caused by incendiary bombs, which rendered the college unusable for several months.[19] After the war, the names of a further twenty-five Westminsterians who were killed between 1939 and 1945 were added to the memorial, which was rededicated during 'a simple ceremony' on 1 December 1946.

Although the premises on Horseferry Road were repaired and reoccupied, the college packed its bags for good in 1959, relocating some 60 miles west to purpose-built facilities in Oxford. The memorial and stained-glass window were also transferred and now stand in the chapel at Harcourt Hill. After considerable debate, the college's former London home was demolished in the 1960s and the land redeveloped.[20] Today, it serves as the headquarters of the British television broadcaster Channel 4, which operates from a Grade II listed building on the site.

The story of Westminster College would be incomplete without acknowledging its other Victoria Cross recipient, William Thomas Forshaw. A native of Barrow-in-Furness, he was a year older than Bell, but the two men were well acquainted, having represented the college together on the track. Lieutenant Forshaw, a company commander with the 1/9th Manchester Regiment, was recognized for his actions during the Gallipoli campaign at the Battle of Krithia Vineyard in August 1915, when his company repelled repeated Turkish counter-attacks over a period of nearly forty-eight hours. He received his VC three months later and survived the war, but died in 1943 at the age of just 53.[21]

On 5 October 1921, a new memorial tablet bearing Don's name was unveiled at St Peter's Church in Harrogate by Henry Lascelles, 5th Earl of Harewood. It commemorates fifty-seven men with connections to the church who died during the war, their names inscribed on two bronze panels flanking a central crucifix. Following a dedication by the Bishop Suffragan of Knaresborough, the choir, in the presence of a 'large congregation', sang the hymns 'Be Thou Faithful Unto Death' and 'Yea, Though I Walk Through the Valley'.[22] The tablet was originally mounted on the west interior wall of the church, but was relocated to its present position, just inside the south-eastern entrance, during extensive redevelopment work carried out in 2011 and 2012.

Harrogatonians had to wait several years for the town's main war memorial to be erected, but few could deny the result justified their patience. Designed by Ernest Prestwich, with bas-relief work by sculptor Gilbert Ledward, more than 10,000 people gathered in Prospect Square on 1 September 1923 to watch the Earl of Harewood unveil the 75ft stone obelisk – described at the time as 'one of the finest in the country and worthy of the town'.[23] The ceremony included a military march-past and the laying of wreaths, with a bugler from the 5th West Yorkshire Regiment sounding the 'Last Post' and 'Reveille' on either side of an impeccably observed two-minute silence.[24] The memorial originally bore the names of 879 local dead from the First World War, inscribed on two large brass plaques. In 1948, an additional 321 names from the Second World War were added, bringing the total to 1,163. Among the dead of both conflicts are several women, including munitions workers, a nurse, members of Queen Mary's Army Auxiliary Corps, the Women's Land Army and the Auxiliary Territorial Service, and a YMCA volunteer.

In 1999, a new memorial plaque mounted on a stone plinth at the foot of the obelisk was unveiled to honour the town's VC recipients. Initially, it carried just two names – those of Donald Bell and Charles Hull. However, in 2007, an additional plaque was erected to commemorate another Harrogate-born VC holder, Sergeant Robert Grant, who was decorated for his bravery at Alumbagh, near Lucknow, on 24 September 1857 during the Indian Mutiny.[25] Flanking the plinth, two memorial paving stones were later installed to remember the actions of the First World War recipients. Hull's was officially unveiled on 5 September 2015, the centenary of his action at Hafiz Kor; Bell's followed ten months later, marking the one-hundredth anniversary of his gallantry at Horseshoe Trench.

Family lore holds that Smith and Annie Bell made a pilgrimage to France in the 1920s to visit their son's grave. Thousands of bereaved families made similar trips, many on official tours organized by the British Legion, founded in 1921. The most notable of these took place in August 1928, when 11,000 ex-servicemen, war widows and relatives embarked on the 'Great Pilgrimage' – a three-day tour of cemeteries and memorials in France and Belgium that culminated in a service of remembrance at the recently unveiled Menin Gate in Ypres on 8 August.[26]

Others made the journey independently, often using commercial travel firms like Thomas Cook & Son, which had offered battlefield tours since 1919, or through charitable organizations such as the St Barnabas Society or the Church Army, which established a network of hostels in converted Nissen huts along the old front line.[27] A plethora of battlefield guidebooks were also available, including some aimed at the relatively new phenomenon of the independent motorist. Among the best known were *The White Cross Touring Atlas of the Western Battlefields* and the *Illustrated Michelin Guides to the Battlefields*, but similar publications were also produced by other companies.

Sadly, the details of any trip made by Don's parents have been lost to time. It is possible that Rhoda accompanied them on their sombre pilgrimage, or that one or more of Don's siblings also made the journey. Unfortunately, we will probably never know. Later generations of the family have certainly made the trip to Gordon Dump Cemetery, preserving the bond and continuing the act of remembrance.

And what of the Bell family after the war? In January 1920, Gertie married Arthur Umpleby, a farmer from nearby Killinghall. The couple had three children and spent much of their lives on their farm at Airton, seven miles north of Skipton. Arthur died in April 1965, but Gertie lived another decade, passing away in 1975 at the age of 87.

By 1921, Smith and Annie Bell were living at 7 Dragon Parade with their son Billy, daughter Dolly and a niece, Winifred Preston. Dolly left home that September, marrying Harold Angus in a ceremony at the Wesley Chapel. Harry was the nephew of the Reverend J.A. Angus and had served with the West Yorkshire Regiment during the war. He was gassed in 1917 and was later sent to the United States, where he trained American troops in Kentucky. The couple had two children, but Harry never fully recovered from the long-term effects of being gassed and died in 1946 from pneumonia.

He was just 49. Dolly outlived her husband by more than thirty years. She died in September 1982, in her eighty-sixth year.

Billy Bell was the last to settle down; he married Edith Thornton at Christ Church in Harrogate in July 1923. Edith was a war widow from Murton in County Durham, whose first husband, Frank – a captain with the 171st Tunnelling Company, Royal Engineers – had been killed in Belgium at the end of 1917.[28] Billy and Edith had one child, a son born in June 1924 and named after his late uncle, Donald Simpson Bell. After the war, Billy worked as a manager with the Harrogate Carriage Company before finding employment as a confectionery salesman.[29] He died at 28 East Parade on 6 May 1962, aged 74. Edith passed away in 1979.

Of the remaining siblings, Nellie lived for many years at 'Somerville' on Anchor Road in High Harrogate with her husband, Harry Jackson. Situated close to the family-owned Jubilee Laundry, the 1939 Register reveals they had two civil servants from the Air Ministry lodging with them when war broke out again. Nellie died in January 1968, aged 82, but Harry lived a further nine years. After his death in 1977, he was buried alongside his wife in the family plot at Stonefall Cemetery on Wetherby Road. Their two sons, named after Nellie's brothers, Billy and Don, are buried with them.

Nancy, who left Britain for South Africa in the summer of 1915, spent the rest of her life in Grahamstown, Eastern Cape Province, where she lived with her husband, Harry Sole. The first of their four children, Donald Bell Sole, was born at the end of 1917, sixteen months after his uncle's death at Contalmaison. Their only daughter, Margaret, was born during a three-month trip back to Harrogate in 1936. Harry, who inherited his father's printing business, died in 1970. Nancy passed away five years later in Port Alfred, in her eightieth year, and was buried beside him in the family plot at Grahamstown, officially known as Makhanda since 2018.[30]

As for Minnie, who had known Don and Rhoda as well as anyone, she lived at 52 Harlow Terrace with her husband, Tom Wood, for the best part of fifty years. After returning from France, Tom worked at Topham Bros Ltd on Oxford Street in Harrogate; he was still listed as a reserve with the Royal Engineers in 1939, despite being in his fifties. Minnie and Tom died within six months of each other in 1968, though both were in their eighties and had enjoyed long and fulfilled lives. They had no children.

Don's death left an aching void in Annie Bell that never truly healed. Like many bereaved mothers, she bore her grief with quiet dignity, but the strain gradually took its toll on her health. On 21 April 1928, 66-year-old Annie passed away at home on Studley Road with her husband at her side.[31] She was buried at Grove Road Cemetery three days later, following a service at Grove Road Wesley Chapel.

After his wife's death, Smith Bell moved to 28 East Parade to live with his son, Billy, daughter-in-law Edith and grandson Donald. He made several trips to South Africa during the 1920s and 1930s to visit Nancy and her family, returning from the last of these in March 1936.[32] Two years later, 79-year-old Smith passed away at a nursing home, and was buried next to Annie in Grove Road Cemetery on 2 May 1938 after a service at the Wesley Chapel. The name of the couple's youngest daughter, Mary, who died in infancy in 1905, is inscribed on the reverse of the grave memorial. Their son Don is commemorated on the front. Sadly, in May 1994, the grave was desecrated by vandals who knocked the cross off the top of the headstone. Although the incident drew widespread condemnation, those responsible were never apprehended.[33]

What became of Rhoda Bell? She returned to her native Kirkby Stephen after the death of her father in 1918, living at her childhood home with her mother, sister, brother-in-law and niece. By the start of the Second World War, the 53-year-old was back in Yorkshire, working as a dressmaker in York, where she lived with her brother, Tom Bonson, and his two daughters, Margaret and Jocelyn.[34] Tom had also endured significant bereavement in his life, losing twin infant sons in 1906 and then his wife, Emily, in 1924.

Rhoda's movements over the following years are more difficult to trace, but she eventually went back to Kirkby Stephen, where she worked first as a solicitor's secretary and later as a librarian. She was highly respected in the town, overseeing the 'distribution of books at the library'[35] for many years. On 28 October 1952, Rhoda died suddenly at home, aged 65. Although her death was unexpected, she had been suffering from cardiovascular disease. Her funeral service took place at Kirkby Stephen Centenary Methodist Church on Friday, 31 October, after which she was buried at the local cemetery. Several months later, the *Penrith Observer* reported that two books, *A Smallholder's Encyclopaedia* and *Historic Farm Houses of Westmorland*, had been presented to the County Library in Kendal in her memory. The donation, made to 'commemorate the work of a former Kirkby Stephen Librarian – the late Mrs Rhoda Bell',[36] reflects the high regard in which she was held by the local community.

Sadly, interest in the First World War declined in the decades following the Second World War. Battlefield pilgrimages had ceased in 1939, and post-1945 Britain was preoccupied with new challenges. The Cold War dominated national discourse, and the pressures of economic recovery, the dismantling of empire and shifting social priorities contributed to the gradual fading of public interest in the war of 1914–18. As its veterans grew older and their numbers diminished, remembrance events were increasingly dominated by the generation that had fought the Second World War.[37]

The late 1960s and 1970s were marked by significant social and cultural change, with growing opposition to nuclear armament and Cold War imperialism. The rise of anti-war and peace movements saw traditional narratives of military sacrifice and remembrance come under intense scrutiny, particularly among younger generations, who increasingly rejected conventional rhetoric and unquestioning reverence for military heroism. Participation in remembrance events declined, while some war memorials became sites of protest and, in some instances, of vandalism.[38]

Harrogate's Wesley Chapel found itself entangled in controversy during these years of political turbulence, when its portrait of Donald Bell was taken down following complaints that it promoted war and militarism. The decision caused considerable distress within the Bell family, particularly for Dorothy, who remained a committed member of the congregation throughout her life. Although the portrait was subsequently reinstated, the episode left a deep and lasting sense of grievance.

The period represented the nadir of battlefield tourism, too. Few people visited the British memorials and cemeteries scattered along the Western Front during the 1960s, 1970s and 1980s. They were maintained with unfailing diligence by the Commonwealth War Graves Commission and remained immaculate, but the absence of pilgrims was evident in the sparse pages of their visitor books. At the most isolated of these 'silent cities', it was not uncommon for a sole signature to stand unaccompanied for many months.

It seemed wholly inconceivable that public interest in the battlegrounds of the Great War would ever be rekindled.[39]

Yet there has been a remarkable upturn in interest over the past couple of decades. Driven by the rise of the internet and the continued digitization of major archival records, historical research is more accessible than ever before. As a result, family historians and genealogists have increasingly sought to discover and document the wartime experiences of their ancestors. This has fuelled a resurgence in battlefield tourism, transforming much of the old front line.[40] Several new museums and visitor centres have opened, larger car parks have been constructed and designated routes drawn up and signposted. A flurry of new memorials have also been erected, while many older ones have been refurbished. This gathered pace as the 2014 centenary of the war approached, but there had already been a steady increase in activity, well before the commemorations.

On 9 July 2000, a new memorial was unveiled near the site of Bell's Redoubt at Contalmaison. Situated on the left-hand side of the road leading out towards Mametz, and just a few hundred yards from Bell's original burial spot, it was funded in large part by the Professional Footballers' Association (PFA), who gave financial support to a project initiated by the late Richard Leake and the Friends of the Green Howards Museum. The memorial was constructed by a local artisan using stone brought from Yorkshire and features a metal replica of the original wooden cross that once marked his grave.[41]

Refurbishment work took place a decade later, followed by a rededication ceremony on 11 September 2010. Members of Don's family were among those present, along with representatives from the PFA and serving officers and men of the Yorkshire Regiment. Ten weeks later, the PFA purchased Bell's Victoria Cross and campaign medals for a hammer price of £221,000 at an auction held by Spink of London.[42] This ensured the medals, previously on loan to the Green Howards Museum, did not pass into the hands of a private collector. They are now on public display at the National Football Museum in Manchester.

There was renewed interest in football's role in the First World War as the centenary approached in 2014. My own groundbreaking research was published that year and featured in the 'Greater Game' exhibition at the National Football Museum. This helped pique interest in Bell's story, and several memorial events were staged in his honour over the next four years. At the 2014 PFA Awards Ceremony, held at Grosvenor House in London, Bell was posthumously awarded the PFA Merit Award, ninety-eight years after his death.[43] This came twelve months after the League Football Education had instituted the Donald Bell VC Award, recognizing players who had overcome adversity during their two-year apprenticeship. He was also honoured by East Coast Trains in 2014, appearing on their specially liveried 'For the Fallen' locomotive, introduced at the start of the centenary commemorations.

Several football matches also took place, including one between two of Bell's former sides, Bradford Park Avenue and Newcastle United, to mark the centenary of his death on 10 July 2016. The match, won 3-0 by the visitors, was played at Bradford's Horsfall Stadium and saw the National League North side debut their new commemorative home shirt for the 2016/17 season, featuring a small tribute to Bell on the reverse.[44] Two months later, the club sent players to France to play alongside members of the Yorkshire Regiment in a friendly against local side Albert, which they won by three goals to two. The match followed a series of commemorative events, including a service at Bell's Redoubt and a wreath-laying ceremony at his grave in Gordon Dump Cemetery. Twenty-eight members of Don's family travelled from the UK, South Africa, New Zealand and Canada for the commemorations, which were also attended by representatives from the Yorkshire Regiment, the Green Howards Association, the PFA and Bradford Park Avenue.[45]

The annual reunion of the Westminster Society also commemorated Bell that year, marking the centenary of the award of his VC with a service in the chapel at Harcourt Hill, where his stained-glass memorial window is now located. Members of the Bell family were in attendance, as were officials from Oxford Brookes University, the PFA, the Yorkshire Regiment and the local Officer Training Corps.[46]

Bell is remembered each year at Harrogate Grammar School's Annual Celebration of Achievement Awards Evening, where two awards are presented in his name: the 'Donald Bell Award for Best School Footballer' and the 'Donald Bell Award for Commitment'. In December 2016, the recipient of the former prize also received a signed commemorative Bradford Park

Avenue shirt. Both awards were presented by Don's great-nephew, Colonel Martin Dransfield of the New Zealand Defence Force, who had travelled from his home in New Zealand for the event.[47]

More than a century after he fell in battle, the story of Donald Simpson Bell remains a source of inspiration. He exemplifies the finest qualities of his generation – courage, selflessness and an unwavering sense of duty. From the classrooms and football pitches of Harrogate to the battlefield where he gave his life, Bell's legacy stands as a lasting testament to extraordinary bravery and sacrifice. One can only imagine the path his life might have taken had war not cut it so tragically short. Asked about his friend shortly before his own death in 1971, Don's former schoolmate and fellow Victoria Cross recipient, Archie White, had this to say:

> 'Probably no one else on the front could have done what he did. Laden with steel helmet, haversack, revolver, ammunition and Mills Bombs in pouches, he was yet able to hurl himself at the German trench at such speed that the enemy would hardly believe what they saw. He was a magnificent soldier; and had he lived, with his high intelligence, superb physique and firm religious principles, he would have risen high in his chosen profession.'[48]

Notes

Prologue

1. *Harrogate Herald*, 22 December 1915.

Chapter 1: Roots and Foundations

1. Grainge, W., *Nidderdale* (Pateley Bridge: Thomas Thorpe, 1863), pp.108–09.
2. *Ibid.*
3. *Ibid.*
4. S. Bell, Birth Certificate, HMRO, 1859.
5. Census Returns of England and Wales, 1881, Class: RG 9; Piece: 3195; Folio: 41; Page: 36; GSU roll: 543093, The National Archives (TNA), Kew; accessed via Ancestry.co.uk.
6. A. Bell, Death Certificate, HMRO, 1866.
7. M.A. Bell, Death Certificate, HMRO, 1874.
8. G.W. Bell, Death Certificate, HMRO, 1874.
9. 1881 Census, R. Bell, Summerbridge.
10. *Ibid.*
11. S. Bell & A. Simpson, Marriage Certificate, HMRO, 1883.
12. Situated at the corner of Beulah Street and Oxford Street.
13. *Yorkshire Evening Post* (*YEP*), 16 January 1931.
14. *YEP*, 15 January 1931.
15. *Leeds Mercury* (*LM*), 4 September 1890.
16. *Ibid.*
17. *Ibid.*
18. E. Simpson, Death Certificate, HMRO, 1894.
19. 1891 Census, D.S. Bell, Harrogate.
20. The Wesley Chapel was designed by architect Henry Francis Lockwood.
21. *Harrogate Herald* (*HH*), 11 August 1915.
22. Hitchen, H.S., Dawson, T.E., & Yeats, J.A., *Celebrating 150 years: The History of Wesley Chapel Harrogate 1862–2012* (Harrogate: Anchorprint, 2012), p.29.
23. The school closed following the Second World War and the building is currently occupied by an Indian restaurant.
24. Census of England and Wales, 1901, County Report: Yorkshire (West Riding).
25. 1901 Census, D.S. Bell, Harrogate.
26. *Knaresborough Post* (*KP*), 15 November 1902.
27. Neesam, Malcolm, *Harrogate Grammar School* (Harrogate: Manor Place Press, 2003), pp.7–9.
28. *HH*, 8 April 1902.
29. *Ibid.*
30. *HH*, 16 September 1903.

31. *Municipal Secondary Day School Log Books*, North Yorkshire Archives.
32. M. Bell, Death Certificate, HMRO, 1905.
33. Aldrich, R., *Education for the Nation: A History of the National System of Education in England and Wales* (London: Cassell, 1996), pp.84–88.
34. White was born in Boroughbridge on 5 October 1891.
35. Letter to Col. J. Forbes, 1970.
36. Thomson, A.A., *The Exquisite Burden* (London: Jenkins Ltd, 1936), pp. 117–18.
37. *Ibid.*
38. University of London Undergraduates Matriculation Results June 1908, p.503.
39. *HH*, 4 April 1917.

Chapter 2: Floreat Westminsteriensis!

1. Pritchard, F.C., *The Story of Westminster Training College 1851–1951* (Westminster: Epworth Press, 1951), pp.13–14.
2. *Ibid.*
3. *Ibid.*
4. Letter to mother, September 1909.
5. *Ibid.*
6. *Sporting Life* (*SL*), 10 April 1907.
7. *Ibid.*
8. *Swindon Advertiser*, 23 July 1909.
9. *Beckenham Journal, Penge and Sydenham Advertiser* (*BJPSA*), 7 August 1909.
10. Westminster Training College Timetable, 1910–11. Oxford Centre for Methodism and Church History (OCMCH).
11. Letter, September 1909.
12. Pritchard, p.112.
13. Principal's letter, 1 July 1909, OCMCH.
14. *Ibid.*
15. Letter, September 1909.
16. *Ibid.*
17. Timetable, OCMCH.
18. Pritchard, pp.13–14.
19. Letter to father, 31 October 1909.
20. Pritchard, F.C., *Westminster Training College*, p.108.
21. *Ibid.*
22. *Ibid.*
23. Smith graduated with a First-Class Honours Degree in Classics in 1901.
24. *North Star* (*NS*), 28 June 1901.
25. *Ibid.*
26. Pritchard, p.110.
27. *SL*, 13 May 1907.
28. Pritchard, F.C., *Westminster Training College*, pp.108–11.
29. *SL*, 17 May 1908.
30. *Ibid.*
31. Pritchard, p. 115.
32. *Ibid.*, pp.107–11.
33. Letter home, 11 October 1909.

34. *Ibid.*
35. *Ibid.*, 25 October 1909.
36. Pritchard, p.125.
37. Not to be confused with Barnet Alston, with whom they merged in 1912.
38. Letter to father, 8 October 1909.
39. Fifth Inter-Year Athletics Sports Programme, OCMCH.
40. *London Daily Chronicle* (*LDC*), 6 May 1910.
41. *SL*, 6 May 1910.
42. Pritchard, p.110.
43. *Ibid.*, 9 May 1910.
44. Westminster Training College, Special Memorial Service Order of Service, OCMCH.
45. *London Evening Standard* (*LES*), 21 May 1910.
46. Letter to mother, June 1911.
47. Westminster College Annual Combined Concert 1911, order of service, OCMCH.
48. Letter to Gertie, 8 February 1911.
49. Letter to Minnie, 21 March 1911.
50. Letter to Gertie, 10 February 1911.
51. *Ibid.*, 15 February 1911.
52. *Ibid.*
53. *SL*, 15 May 1911.
54. Logbook, May 1911, OCMCH.
55. *The Sportsman* (*SPO*), 15 May 1911.
56. Logbook June 1911, OCMCH.

Chapter 3: A Man of Two Callings

1. *HH*, 25 July 1911.
2. *Ibid.*
3. *Ibid.*
4. *Ibid.*
5. *YEP*, 26 July 1911.
6. Starbeck Council School Logbook, 24 July 1911, North Yorkshire Archives.
7. Don's letter to Gertie, April 1911.
8. It was renamed Harrogate Grammar School in 1931 and moved to its present site in 1933.
9. Starbeck Council School Logbook.
10. Joannou, P., *The First 100 Years: The Official History of Newcastle United Football Club 1882–1982* (Newcastle: Newcastle United Football Company, 1984).
11. *Northern Echo* (*NE*), 15 August 1911.
12. *Sheffield Daily Telegraph* (*SDT*), 21 August 1911.
13. *Sports Argus* (*SA*), 26 August 1911.
14. McCracken and Hudspeth both spent nineteen seasons at the club.
15. Betts joined Derby in October 1911.
16. *Liverpool Evening Express* (*LEE*), 21 October 1911.
17. Joannou, P., *The Black 'n' White Alphabet* (Leicester: Polar Publishing, 1996), p.176.
18. *Football Gazette and Telegraph* (*FG*), 2 September 1911.
19. *Newcastle Chronicle's Football Gazette* (*NC*), 2 September 1911.
20. *NC*, 30 September 1911.

21. *NS*, 16 October 1911.
22. *Lancashire Evening Post* (*LEP*), 23 October 1911.
23. *NS*, 2 October 1911.
24. *Yorkshire Post* (*YP*), 13 November 1911.
25. *Durham Chronicle* (*DUR*), 24 November 1911.
26. *Ripon Observer* (*RO*), 23 November 1911.
27. *Ibid.*
28. Donald Bell Jackson died in 2009, aged 85.
29. *YEP*, 30 December 1911.
30. Bishop Auckland actually toured Hungary and Romania.
31. *FG*, 6 January 1912.
32. *Newcastle Daily Chronicle* (*NDC*), 5 January 1912.
33. *FG*, 6 January 1912.
34. *FE*, 20 January 1912
35. *NC*, 24 February 1912.
36. *FG*, 23 March 1912.
37. *RO*, 11 April 1912.
38. *Ibid.*
39. *Ibid.*
40. *Ibid.*
41. *YP*, 10 April 1912.
42. *LM*, 15 April 1912.
43. *Halifax Evening Courier* (*HEC*), 15 April 1912.
44. *Ibid.*
45. *LM*, 15 April 1912.
46. Lord, W., A *Night to Remember* (London: Penguin Books, 2002).
47. *NDC*, 29 April 1912.
48. *Athletic News* (*AN*), 15 January 1912.
49. *YP*, 30 April 1912.
50. *AN*, 26 August 1912.
51. *DC*, 28 August 1912.
52. *Wharfdale & Airdale Observer*, 20 September 1912.

Chapter 4: Up the Avenue!

1. Inglis, S., *Soccer in the Dock* (London: Willow, 1985), pp.12–13.
2. *Ibid.*, p.18.
3. *AN*, 18 June 1906.
4. *The Times* (*TT*), 15 July 1910.
5. *The Football Echo*, 24 December 1910.
6. Ward and Maddocks, pp.62–65.
7. Williams, G., *The Code War: English Football Under the Historical Spotlight* (London: Mainstream Publishing, 2001), pp.98–101.
8. Hartley, M. and Clapham, T., *The Avenue* (Nottingham: Temple Nostalgia, 1987).
9. Smith had scored forty-nine goals in fifty-nine games for Brighton. He joined Bradford for a fee of £735 plus inside forward Bobby Simpson.
10. *YP*, 19 October 1912.
11. *FG*, 26 October 1912.

12. *YP*, 7 April 1913.
13. *YEP*, 4 January 1913.
14. *Hull Daily Mail* (*HDM*), 14 April 1913.
15. *TT*, 3 February 1913.
16. *LM*, 17 April 1913.
17. *LM*, 17 April 1913.
18. *YEP*, 16 April 1913.
19. *Ibid.*, 17 April 1913.
20. *SL*, 17 April 1913.
21. *Bradford Daily Telegraph* (*BDT*), 16 April 1913.
22. *YP*, 21 April 1913.
23. *Ibid.*, 27 April 1913.
24. *RO*, 17 April 1913.
25. *YEP*, 2 August 1913.
26. *LM*, 22 August 1913.
27. *BDT*, 28 August 1913.
28. Also known as the West Yorkshire Cup at the time.
29. *YP*, 2 October 1913.
30. *Ibid.*
31. *Ibid.*
32. *The Park Avenue Journal*, 11 October 1913.
33. *BDT*, 2 October 1913.
34. *Ibid.*, 4 October 1913.
35. *Ibid.*, 13 October 1913.
36. *LM*, 20 October 1913.
37. *YP*, 20 October 1913.
38. *Ibid.*
39. *BDT*, 21 October 1913.
40. *Football News* (*FN*), 25 October 1913.
41. *BDT*, Monday 27 October 1913.
42. *Ibid.*
43. *Ibid.*
44. *LM*, 27 October 1913.
45. *LEP*, 27 October 1913.
46. *Ibid.*
47. *BDT*, 3 November 1913.
48. *Leicester Daily Post* (*LEP*), 3 November 1913.
49. *Ibid.*
50. *Ibid.*
51. *Ibid.*, 8 November 1913.
52. *BDT*, 8 November 1913.
53. *Ibid.*, 7 November 1913.
54. *YP*, 10 November 1913.
55. *BDT*, 10 November 1913.
56. *YP*, 10 November 1913.
57. *Ibid.*
58. *AN*, 10 November 1913.

59. *BDT*, 10 November 1913.
60. *BDT*, 17 November 1913.
61. *Ibid.*, 24 November 1913.
62. *Ibid.*, 13 December 1913.
63. *Barrow Herald*, 9 September 1913.
64. *BDT*, 14 September 1913.
65. *Ibid.*, 14 November 1913.
66. Mirfield United were one of five teams who could not complete all of their fixtures. However, they could not catch Bradford at the top, even if they had played their one outstanding fixture.
67. *DM*, 25 April 1913.
68. *Ibid.*
69. *BDT*, 27 April 1914.
70. *Ibid.*
71. *Ibid.*
72. *Ibid.*, 27 April 1914.
73. *Ibid.*
74. *Census of England and Wales, 1911: General Report with Appendices* (London: His Majesty's Stationery Office, 1917), pp.12–45.
75. Dewhirst, J., *Life at the Top* (Shipley: Bantamspast, p.208).
76. *Ibid.*

Chapter 5: Controversy and Condemnation

1. Gregory, A., *The Last Great War: British Society and the First World War* (Cambridge: Cambridge University Press, 2008), p.73.
2. *Manchester Evening News* (*MEN*), 7 August 1914.
3. *Newcastle Journal* (*NJ*), 20 August 1914.
4. *TS*, 27 August 1914.
5. Broom, T., *Cricket in the First World War: Play Up! Play the Game* (Cheltenham: The History Press, 2021).
6. Collins, T., 'Rugby League in World War One', *Rugby Reloaded* (5 August 2014), accessed 4 June 2024, https://tony-collins.squarespace.com/rugbyreloaded/2014/8/5/rugby-league-in-world-war-one.
7. *TT*, 15 August 1914.
8. *Official Programme: England v France, Twickenham, 13 February 1915* (London: Rugby Football Union, 1915). Includes a statement encouraging rugby players to enlist.
9. Sir Arthur Conan Doyle speech, 6 September 1914.
10. *LDP*, 26 November 1914.
11. Charrington, F.N., *Football and the War* (London: 1914).
12. Cavallini, R., *Play Up Corinth: A History of the Corinthian Football Club* (Chalford: Stadia, 2007).
13. *SL*, 28 August 1914.
14. *BDT*, 27 August 1914.
15. *DM*, 31 August 1914.
16. *TS*, 1 September 1914.
17. *MEN*, 31 August 1914.
18. *BDT*, 7 September 1914.

19. *Ibid.*
20. *YEP*, 12 September 1914.
21. Simon J., 'A Different Kind of Test Match: Cricket, English Society and the First World War', *Sport in History*, 33, no. 1 (2013), pp.21–26.
22. *TS*, 7 September 1914.
23. *TIM*, 8 September 1914.
24. *Sunderland Daily Echo* (*SDE*), 2 September 1914.
25. *Western Mail*, 1 September 1914.
26. *TS*, 14 September 1914.
27. Riddoch, A., *When the Whistle Blows: The Story of the Footballers' Battalion in the Great War* (Stroud: Haynes Publishing, 2008), p.16.
28. Harding, J., *Behind the Glory: 100 Years of the PFA* (Derby: Breedon, 2009), p.93.
29. *Ibid.*
30. *BDT*, 4 September 1914.
31. *Ibid.*
32. *LDC*, 8 October 1914.
33. Walden had pre-war service with the Cheshire Regiment and would later join the 18th W. Yorks.
34. Jenkins, S., *They Took the Lead* (Chippenham: Yore Publications, 2004), pp.56–58.
35. *The Courier* (*COU*), 5 September 1914.
36. *Sheffield Green 'Un* (*SGU*), 12 September 1914.
37. *SC*, 26 July 1916.
38. Corporal Joseph Stanislaus Maley died of wounds on 17 May 1915 and is buried at Béthune Town Cemetery.
39. *LEP*, 19 July 1915.
40. *BDT*, 2 October 1914.
41. Leake, R., *A Breed Apart* (Scarborough: Great Northern), p.25.
42. TNA, WO 363.
43. *LM*, 19 October 1914.
44. Private 15722 D.S. Bell, 'A' Company, 9th (Service) Battalion, The Prince of Wales's Own (West Yorkshire Regiment).
45. Billy letter to mother, undated.
46. *LDC*, 24 November 1914.
47. *Pall Mall Gazette* (*PMG*), 28 November 1914.
48. *YEP*, 28 November 1914.
49. The 16th (Service) Battalion, Royal Scots (Lothian Regiment) was raised by Sir George McCrae and contained a number of Scottish players, including sixteen from Heart of Midlothian.
50. Riddoch, A., pp.30–36.
51. *Ibid.*
52. *The Scotsman*, 30 March 1915.
53. *LDP*, 26 April 1915.
54. *LEP*, 26 April 1915.
55. *Ibid.*, 20 July 1915.
56. *Ibid.*, 30 July 1915.
57. *Birmingham Daily Post*, 27 July 1915.
58. *Ibid.*

59. *LEP*, 20 July 1915.
60. TNA, WO 372/12/19840.
61. *LM*, 15 November 1915.
62. TNA, WO 372/18/135104.
63. *Liverpool Echo* (*LE*), 20 July 1915.

Chapter 6: King's Regulations

1. Whitehouse, C.J. & C.P., *A Town for Four Winters* (Brocton: Whitehouse, 1978), p.5.
2. Officially known as the 11th (Reserve) Battalion, Alexandra, Princess of Wales's Own (Yorkshire Regiment). Also known as the 'Green Howards'.
3. 1911 Census, R. Bonson, Wilmslow.
4. National Trust, Belton House and the First World War, National Trust, 2025. Accessed 28 April 2025, https://www.nationaltrust.org.uk/visit/nottinghamshire-lincolnshire/belton-estate/belton-house-and-the-first-world-war.
5. White letter to Col. J. Forbes, 1970.
6. *The London Gazette* (*LG*). Supplement 29181. 8 June 1915.
7. *NDC*, 16 June 1915.
8. Letter to mother, 10 May 1916.
9. Trapmann, A.H., *Straight Tips for 'Subs'* (London: Forster Groom Ltd, 1915), pp.15–19.
10. Billy letter to mother, 10 June 1915.
11. *HH*, 11 August 1915.
12. Messenger, C., *Call to Arms: The British Army 1914–18* (London: Weidenfeld & Nicolson, 2005), p.300.
13. Trapmann, p.25.
14. Trapmann, p.28.
15. Messenger, p.309.
16. United Kingdom, Military Service Act 1916. 5 & 6 Geo. 5 c. 104. London: HMSO, 1916.
17. Messenger, pp.314–34.
18. *Ibid.*
19. Wylly, H.C., *The Green Howards in the Great War* (Richmond: Green Howards Regimental Headquarters, 1926), pp.392–93.
20. Billy letter to mother, October 1915.
21. *Ibid.*
22. *LG*, 29 June 1915.
23. Letter home, 24 November 1915.
24. Holmes, R., *The Oxford Companion to Military History* (Oxford: Oxford University Press, 2001), s.v. 'batman'.
25. Letter home, 24 November 1915.
26. *Ibid.*
27. *Ibid.*
28. Pritchard, pp.121–25.
29. 'Records Relating to Wartime Relocation and Operations, 1914–1918' (Richmond: Wesleyan Theological College Archives).
30. Long, E.A., *Folkestone in the Great War* (Barnsley: Pen & Sword Books, 2014).
31. Bagwell, P., *The Railwaymen* (London: George Allen & Unwin, 1963), p.211.
32. Letter home, 24 November 1915.

33. Easdown, M., *The History of the Metropole Hotel*(Folkestone: Folkestone & District Local History Society, 2004), p.2.
34. *Ibid.*
35. *Ibid.*
36. Letter home, 24 November 1915.
37. Long, p.15.
38. Douie, C.W., *The Weary Road: Recollections of a Subaltern of Infantry* (London: Heinemann, 1929), p.37.
39. Dunn, S., *Securing the Narrow Sea: The Dover Patrol 1914–1918* (Barnsley: Seaforth Publishing, 2017), pp.5–10.
40. Reay, S., *The Half-Shilling Curate: A Personal Account of War & Faith 1914–1918* (Solihull: Helion & Company, 2016), pp.101–12.
41. Plowman, M., *A Subaltern on the Somme* (London: T. Fisher Unwin, 1927), p.9.
42. *Ibid.*, p.9.
43. Reay, pp.59–61.
44. Owen, W., Letter to his mother, Étaples, France, 1917. *Letters of Wilfred Owen,* ed. Harold Owen (London: Oxford University Press, 1967), p.505.
45. The War Diary of Robert Lindsay MacKay.
46. Blunden, E., *Undertones of War* (London: R. Cobden-Sanderson, 1928), p.17.
47. Letter home, 26 November 1915.
48. Wylly, pp.176–81.
49. Letter home, 24 November 1915.
50. Douie, pp.38–43.
51. Plowman, pp.38–40.
52. Douie, pp.38–43.
53. The War Diary of George Culpitt, Royal Welch Fusiliers.
54. *Ackrill's Harrogate War Souvenir*, p.76.
55. Plowman, p.14.
56. Thomas, A., *A Life Apart* (London: Victor Gollancz, 1968), p.44.
57. Yapp, A.K., *The Romance of the Red Triangle* (London: Hodder & Stoughton, 1918), p.123.
58. *Ibid.*, pp.146.
59. Gill, D. & Dallas, G., 'Mutiny at Étaples Base in 1917', *Past & Present*, no. 69 (November 1975), pp.88–112.
60. Carrington, C., *Soldier from the Wars Returning* (London: Hutchinson, 1965), p.245.
61. *Ibid.*
62. Holmes, R. *Tommy: The British Soldier on the Western Front 1914–1918* (London: Harper Collins, 2004), p.395.
63. Letter to mother, 23 November 1915.

Chapter 7: Nursery Sector

1. Herbin, P., *Erquinghem-Lys – Le Fort Rompu* (ADLFI, Archéologie de la France, 2017).
2. Edmonds, J.E., *Military Operations France and Belgium, 1914: October–November 1914* (London: Macmillan, 1925).
3. Letter to father, 29 November 1915.
4. *Huddersfield Daily Examiner* (*HDE*), 21 September 1914.

5. Sheen, J., *The Green Howards in the Great War* (Barnsley: Pen & Sword Military, 2024), pp.8–13.
6. *Surrey Advertiser*, 19 September 1914.
7. 'War Diary of the 9th (Service) Battalion, Alexandra, Princess of Wales's Own (Yorkshire Regiment)', WO 95/2184/3.
8. *Ibid.*
9. *Ibid.*
10. *Ibid.*
11. *Ibid.*
12. *Ibid.*
13. Letter to mother, 1 December 1915.
14. 'War Diary of the 69th Infantry Brigade', WO 95/2183, TNA, Kew.
15. *Ibid.*
16. War Office, *The Training and Employment of Grenadiers* (London: His Majesty's Stationery Office, 1915).
17. *Ibid.*
18. *Ibid.*
19. *Ibid.*, p.3.
20. *Ibid.*
21. *Ibid.*
22. *Ibid.*
23. Letter to Gertie, 29 November 1915.
24. *Ibid.*
25. Letter home, 29 November 1915.
26. *Ibid.*
27. WO 95/2184/3.
28. *TT*, 17 December 1915.
29. *Ibid.*, 14 May 1915.
30. *Ibid.*, 19 December 1915.
31. WO 95/2184/3.
32. Edmonds, *Military Operations, 1915*.
33. *Ibid.*
34. *HH*, 22 December 1915.
35. WO 95/2184/3.
36. *HH*, 22 December 1915.
37. Letter to father, 26 November 1915.
38. *Ibid.*, 16 December 1915.
39. Letter to mother, 22 December 1915.
40. *Ibid.*
41. *Ibid.*
42. *Ibid.*, 12 January 1916.
43. *Ibid.*

Chapter 8: All Roads Lead to the Somme

1. TNA, WO 95/2184/3.
2. Wylly, pp.291–93.
3. *Ibid.*

4. Sandilands, H.R., *The 23rd Division, 1914–1919* (Edinburgh and London: William Blackwood and Sons, 1925), pp.41–42.
5. *Ibid.*
6. *Ibid.*
7. TNA, WO 95/2167/2.
8. Sandilands, *The 23rd Division*, pp.41–42.
9. Letter to mother, 5 January 1916.
10. Letter to mother, 15 January 1916.
11. *Ibid.*
12. TNA, WO 95/2184/3.
13. TNA, WO 95/2183.
14. Letter to Minnie, 16 February 1916.
15. Plowman, p.12.
16. Letter to mother, 19 February 1916.
17. TNA, WO 95/2183.
18. Letter to mother, 1 March 1916.
19. Dodgson was killed during the advance towards Contalmaison on 10 July 1916.
20. Fair, C., *Marjorie's War: Four Families in the Great War 1914–1918* (London: Ebury Press, 2013), p.130.
21. Letter to mother, 9 March 1916.
22. *Ibid.*
23. *Ibid.*
24. TNA, WO 95/2183.
25. Fair, *Marjorie's War*, p.130.
26. Sandilands, *The 23rd Division*, p.48.
27. *Ibid.*, p.250.
28. Sumner, Ian, *They Shall Not Pass: The French Army on the Western Front 1914–1918* (Barnsley: Pen & Sword Military, 2012).
29. *Ibid.*
30. *LG*, 3 March 1916.
31. Letter to mother, 16 March 1916.
32. *Shields Daily Gazette*, 15 March 1916.
33. TNA, WO 95/2184/3.
34. Letter to mother, 16 March 1916.
35. *Ibid.*
36. Letter to mother, 29 March 1916.
37. Rawson, pp.245–46.
38. Letter to mother, 29 March 1916.
39. TNA, WO 95/2183.
40. Letter to mother, 9 April 1916.
41. *Ibid.*
42. Letter to mother, 20 April 1916.
43. *LM*, 13 May 1916.
44. Letter to Dolly, 19 May 1916.
45. 2/Lt J.M. Simpson, 173rd Tunnelling Coy, RE, killed on 9 May 1916 and buried at Nœux-les-Mines Communal Cemetery.
46. Billy letter to mother, 28 May 1916.

47. *HH*, 7 June 1916.
48. *YP*, 20 September 1918.
49. Messenger, C., *Call to Arms: The British Army 1914–18* (London: Weidenfeld & Nicolson, 2005), pp.130–69.
50. *Ibid.*
51. Letter to Minnie, 12 April 1916.
52. Letter to mother, 27 April 1916.
53. *Ibid.*, 28 May 1916.
54. Campbell, J., *Jutland: An Analysis of the Fighting* (London: Conway Maritime Press, 1998).
55. *HH*, 7 June 1916.
56. *Ibid.*
57. *DM*, 5 June 1966.
58. *Ibid.*
59. Sandilands, *The 23*rd *Division*, p.61.
60. Unpublished manuscript, Green Howards Museum.
61. *Ibid.*
62. Letter to mother, 29 June 1916.

Chapter 9: The Mouth of Hell

1. Macdonald, L., *Somme* (London: Penguin, 1983), p.75.
2. *Ibid.*
3. *Ibid.*, p.80.
4. Middlebrook, M., *The First Day on the Somme: 1 July 1916* (London: Allen Lane, 1971), pp.128–29.
5. Sheffield, G., *The Somme: A New History* (London: Cassell, 2003), p.84.
6. *Ibid.*
7. *Ibid.*
8. Letter to mother, 29 June 1916.
9. WO 95/2183.
10. Wylly, pp.291–93.
11. Edmonds, *Military Operations: France and Belgium, 1916, Vol. I*, pp.285–393.
12. *Ibid.*
13. Becke, A.F., *History of the 8*th *Division 1914–1918* (Aldershot: Gale & Polden, 1922), p.73.
14. TNA WO 95/2184/3.
15. Wylly, p.293.
16. *Ibid.*
17. TNA WO 95/2183.
18. *Ibid.*
19. Fussell, P., *Myth, Ritual, and Romance. The Great War and Modern Memory* (Oxford: Oxford University Press, 2013), pp.131–35.
20. *Ibid.*
21. Edmonds, *Military Operations: France and Belgium, 1916, Vol. I*, pp.375–85.
22. *Ibid.*, p.370
23. *Ibid.*, p.576.
24. *Ibid.*, p.379.

25. *Ibid.*, pp.375–76.
26. TNA WO 95/2183.
27. Sheen, J., *Tyneside Irish: 24*th*, 25*th*, 26*th *and 27*th *(Service) Battalions of the Northumberland Fusiliers* (Barnsley: Pen & Sword Books, 2010), pp.93–109.
28. Edmonds, *Military Operations: France and Belgium, 1916, Vol. I*, pp.375–76.
29. TNA WO 95/2183.
30. Sandilands, *The 23*rd *Division*, p.65.
31. TNA WO 95/2183.
32. *Ibid.*
33. *Ibid.*
34. *Ibid.*
35. *Ibid.*
36. Edmonds, *Military Operations: France and Belgium, 1916, Vol. I*, p.191.
37. Sandilands, *The 23*rd *Division*, p.66.
38. War Diary of the 11th West Yorkshire Regiment, TNA WO 95/1809/2.
39. TNA WO 95/2184.
40. *Ibid.*
41. *Ibid.*
42. *Ibid.*
43. *Ibid.*
44. Sandilands, *The 23*rd *Division*, p.67.
45. TNA WO 95/2183.
46. *Ibid.*
47. TNA WO 95/2183.
48. *Ibid.*
49. *Ibid.*
50. Sandilands, *The 23*rd *Division*, p.66.
51. TNA WO 95/2183.
52. TNA WO 95/2183.
53. TNA WO 95/2183.
54. Sandilands, *The 23*rd *Division*, p.67.
55. *Ibid.*
56. *Ibid.*

Chapter 10: These Stirring Times

1. *Ibid.*
2. TNA WO 95/2183.
3. *Ibid.*
4. Wylly, p.296.
5. *Ibid.*, p.294.
6. *Marlborough College at War* (Marlborough: Marlborough College, n.d.), pp.78–79.
7. Wylly, p.296.
8. *LM*, 13 September 1916.
9. Letter to mother, 7 July 1916.
10. TNA WO 95/2183.
11. *The Green Howards Gazette*, November 1924, p.134.
12. *Ibid.*

13. *Ibid.*
14. *Ibid.*
15. TNA WO 95/2184/3.
16. Letter from an orderly of Captain John Cecil Rix, RAMC.
17. TNA WO 95/2184/3.
18. Wylly, p.296.
19. TNA WO 95/2183.
20. Letter to mother, 7 July 1916.
21. Letter to Nancy, 7 July 1916.
22. *The Harrogatonian*, December 1915, p.4.
23. TNA WO 95/2184/3.
24. *HH*, 19 July 1916.
25. *Ibid.*
26. TNA WO 95/2183.
27. TNA WO 95/2184/3.
28. TNA WO 95 95/1717/1.
29. TNA WO 95/2184/3.
30. TNA WO 95/2184/2.

Chapter 11: A Tiny French Hamlet

1. Service du Recensement Général, Recensement général de la population: Département de la Somme, 1911 (Paris: Ministère du Travail et de la Prévoyance sociale, 1911).
2. Masefield, J., *The Battle of the Somme* (London: William Heinemann, 1919), pp.36–37.
3. O'Mara, D., *The French on the Somme: From Serre to the River Somme* (Barnsley: Pen & Sword Military, 2018), pp.37–39.
4. *Ibid.*
5. Wylly, pp.297–99.
6. Miles, W., *Military Operations: France and Belgium, 1916, Volume II:* 2nd *July 1916 to the End of the Battles of the Somme* (London: Macmillan, 1938), pp.54–59.
7. *Ibid.*, pp.248–53.
8. *Ibid.*, p.377..
9. *Ibid.*, p.377.
10. Shakespear, J., *The Thirty-Fourth Division* (Uckfield: Naval & Military Press, 2001), pp.35–65.
11. Senior Lieutenant Kienitz, Machine Gun Company, 110th Reserve Infantry Regiment.
12. Alexander, J., *McCrae's Battalion: The Story of the 16*th *Royal Scots* (Edinburgh: Mainstream Publishing, 2003), p.167.
13. Miles, *Military Operations: France and Belgium, 1916, Vol. II*, p.299.
14. *Ibid.*
15. Shakespear, *The Thirty-Fourth Division*, pp.35–65.
16. Edmonds, *Military Operations: France and Belgium, 1916, Vol. II*, p.43.
17. TNA WO 95/2184/3.
18. *Ibid.*
19. *Ibid.*
20. TNA WO 95/2183.
21. *Ibid.*
22. TNA WO 95/2184/3.

23. *Ibid.*
24. TNA WO 95/2183.
25. Gibbs, P., *The Battles of the Somme* (London: Heinemann, 1917), pp.80–83.
26. von Stosch, A., *Somme-Nord. Teil I: Die Brennpunkte der Schlacht im Juli 1916* (Berlin: Gerhard Stalling, 1927), p.206.
27. TNA WO 95/2183.
28. *Ibid.*
29. Gibbs, *The Germans on the Somme*, pp.11–12.
30. TNA WO 95/2183.
31. von Stosch, *Somme-Nord*, p.206.
32. *Ibid.*
33. Fuller, J.F.C., 'The Other Side of the Hill', *The Army Quarterly*, IX (1 & 2), October 1924 and January 1925, pp 245–59.
34. *Ibid.*
35. *Civil & Military Gazette*, 29 July 1916.
36. Great Britain, War Office, *Statistics of the Military Effort of the British Empire during the Great War, 1914–1920* (London: His Majesty's Stationery Office, 1922).
37. *Ibid.*
38. TNA WO 95/2170-2.
39. TNA WO 95/2183.

Chapter 12: Into the Breach

1. TNA WO 95/2183.
2. *Ibid.*
3. *Ibid.*
4. Wylly, p.297.
5. TNA WO 95/2183.
6. Miles, *Military Operations: France and Belgium, 1916, Vol. II*, p.56.
7. TNA WO 95/2183.
8. *Ibid.*
9. Fuller, *The Army Quarterly*, IX, pp.245–59.
10. *Ibid.*
11. *Ibid.*, p.256.
12. Sandilands, *The 23*rd *Division*, p.76.
13. TNA WO 95/2183.
14. Miles, *Military Operations*, p.56.
15. Fuller, *The Army Quarterly*, IX, pp.245–59.
16. Miles, *Military Operations*, p.56.
17. TNA WO 95/2183.
18. Miles, *Military Operations*, p.55.
19. *Ibid.*
20. Sandilands, *The 23*rd *Division*, p.78.
21. TNA WO 95/2183.
22. Sandilands, *The 23*rd *Division*, p.77.
23. Miles, *Military Operations*, p.56.
24. TNA WO 95/2183.
25. *Ibid.*

26. Gibbs, *The Germans on the Somme*, pp.11–12.
27. Wylly, p.255.
28. Edmonds, *Military Operations: France and Belgium, 1916, Vol. II*, p.58.
29. TNA WO 95/2183.
30. Edmonds, p.56.
31. *Ibid.*
32. Gibbs, *The Germans on the Somme*, pp.11–12.
33. *Birmingham Daily Mail*, Thursday, 13 July 1916.
34. Miles, *Military Operations*, p.56.
35. Gosse, P., *Memoirs of a Camp Follower* (London: William Heinemann, 1934), p.92.
36. TNA WO 95/2183.
37. *Ibid.*
38. *Ibid.*
39. Miles, *Military Operations*, p.57.
40. Sandilands, p.80.
41. *Ibid.*
42. *Ibid.*
43. *Ibid.*
44. Sandilands, *The 23*rd *Division*, p.80.
45. *Ibid.*
46. Wylly, p.255.
47. *LM*, 13 September 1916.
48. *Ibid.*
49. *Ibid.*
50. *Ibid.*
51. TNA WO 95/2184.
52. *HH*, 9 May 1917.
53. TNA WO 95/2184.
54. Sgt 9113 J. Whitely MM, Pte 16740 J. Devlin of the 8th Yorks and 2/Lt J.D. Marks of the 10th Duke of Wellingtons.
55. TNA WO 95/2184.
56. *Ibid.*
57. *Ibid.*
58. *LDP*, 14 July 1916, p.4.
59. TNA WO 95/2184.
60. Wylly, pp.299–300.
61. Commonwealth War Graves Commission.
62. TNA WO 95/2184/2.
63. *Ibid.*
64. Wylly, p.300.

Chapter 13: A Great Loss to the Battalion

1. Edmonds, *Military Operations: France and Belgium, 1916, Vol. I*, see Appendix III, pp.576–80, for detailed casualty figures for 1 July 1916.
2. Middlebrook, *The First Day on the Somme*, pp.262–72.
3. Miles, *Military Operations: France and Belgium, 1916, Vol. II*, pp.62–88.

4. *Statistics of the Military Effort of the British Empire during the Great War 1914–1920* (HMSO, 1922).
5. *Ibid.*
6. Leighton, M., *Boy of My Heart, 1916* (London: Hodder and Stoughton, 1916), p.220.
7. TNA WO 339/2902 Army Service Record of 2/Lt D.S. Bell.
8. Smith Bell letter to War Office, TNA WO 339/2903.
9. Rhoda Bell letter to War Office, TNA WO 339/2903.
10. Hetherington, A., *British Widows of the First World War: The Forgotten Legion* (Barnsley: Pen & Sword, 2018), p.21.
11. *Ibid.*
12. *Ibid.*, pp.21–23.
13. Letter to mother, 10 May 1916.
14. Private John Byers letter to Rhoda Bell, 14 November 1916, family archives.
15. *Ibid.*
16. *Ibid.*
17. Harry Angus letter, undated.
18. *The Weekender*, 29 October 1982.
19. *Ibid.*
20. Westminster Training College Monthly War Bulletin, OCMCH, p.561.
21. Pritchard, *Westminster Training College*, pp.123–35.
22. *BDT*, 18 July 1916.
23. *SC*, 26 July 1916.
24. *Ibid.*
25. Billy Bell diary entry, 19 July 1916.
26. *HH*, 26 July 1916.
27. *The Daily Sketch*, 15 September 1916.
28. *NJ*, 11 November 1916.
29. *HH*, 20 September 1916.
30. *Ibid.*, 2 August 1916.
31. *Ibid.*, 2 August 1916.
32. *Ibid.*, 30 August 1916.
33. *Ibid.*, 10 January 1917.
34. TNA WO 339/2903.
35. *Ibid.*

Chapter 14: The Most Distinguished and Most Coveted of All

1. *LG*, 8 September 1916.
2. *Ibid.*
3. *Ibid.*
4. *HH*, 13 September 1916.
5. TNA WO 95/2183.
6. TNA WO 339/2902.
7. Hitchen, H.S. *et al.*, *Celebrating 150 years: The History of Wesley Chapel Harrogate 1862–2012*, p.35.
8. *HH*, 20 September 1916.
9. *DC*, 15 December 1916.
10. TNA WO 363, Find My Past.

11. Letter to mother, 7 July 1916.
12. TNA WO 363, British Army Service Record of Private 16748 Joseph Batey. Ancestry.
13. *BDT*, 22 September 1916.
14. *SA*, 23 September 1916.
15. *HH*, 20 September 1916.
16. *Ibid.*
17. *WTC War Bulletin*, October 1916, pp.16–17.
18. *LM*, 16 December 1916.
19. *DM*, 14 December 1916.
20. *NJ*, 14 December 1916.
21. *Ibid.*
22. Billy letter to mother, 25 December 1916.
23. *Ibid.*, 28 May 1917.
24. Billy diary, 5 January 1917.
25. Billy letter to mother, 14 January 1917.
26. *HH*, 10 January 1917, p.4.
27. *Ibid.*, 14 February 1917.
28. *LM*, 31 March 1917.
29. *Ibid.*
30. *HH*, 4 April 1917.
31. *Ibid.*
32. *LM*, 17 May 1918.
33. Edmonds, *Military Operations, France and Belgium, 1917, Vol I*, pp.1–50.
34. *Ibid.*
35. Billy letter to mother, 20 March 1917.
36. *HH*, 27 June 1917.
37. Billy diary entry, 20 July 1917.
38. Billy letter to mother, 21 July 1917.
39. *Ibid.*, 21 October 1917.
40. *Ibid.*, 3 February 1918.
41. *Ibid.*, 17 March 1918.
42. *Ibid.*, 1 April 1918.
43. Edmonds, *Military Ops, 1918, Vol I*, pp.476–81.
44. *Ibid.*, pp.1–50.
45. *Ibid.*
46. *Ibid.*, pp.1–45.
47. *Ibid.*, p.6.
48. *Ibid.*
49. *Ibid.*, pp.1–400.
50. Sheffield, p.158.
51. *LM*, 15 October 1917.
52. *AN*, October 1918.
53. *LG*, 14 December 1918.
54. Douie, p.16.
55. Billy letter to mother, 12 November 1918.
56. Billy service records.

Chapter 15: Epilogue

1. Longworth, P., *The Unending Vigil: A History of the Commonwealth War Graves Commission 1917–1984* (London: Leo Cooper, 1985), pp.86–88.
2. CWGC.
3. *Ibid.*
4. Crane, D., *Empires of the Dead: How One Man's Vision Led to the Creation of WWI's War Graves* (London: William Collins, 2013), pp.93–95.
5. Longworth, *The Unending Vigil*, pp.86–90.
6. King, A., *Memorials of the Great War in Britain: The Symbolism and Politics of Remembrance* (Oxford: Berg, 1998), pp.20–23.
7. Hally, M., '19th July 1919 Peace Day in Britain', *The Western Front Association*, Bulletin 114, August 2019.
8. *TT*, 21 July 1919.
9. Crane, *Empires of the Dead*, pp.147–51.
10. Hanson, N., *The Unknown Soldier: The Story of the Missing of the First World War* (London: Doubleday, 2005), pp.349–62.
11. *HH*, 30 June 1920.
12. *Ibid.*
13. Hitchen, H,S. *et al.*, *Celebrating 150 Years: The History of Wesley Chapel Harrogate 1862–2012*, pp.42.
14. *Ibid.*, pp.42–44.
15. *Wesley Chapel Memorial Records*, *c.* 1920; Wesley Sunday School Plaque, n.d.
16. Pritchard, *The Story of Westminster Training College*, pp.13–14, 134–35.
17. *Ibid.*
18. *Ibid.*
19. Memorial Window, Trust Fund, and Bomb Damage Reports (Oxford: Westminster College Special Collections), pp.182–86.
20. *Ibid.*
21. *LG*, 7 September 1915.
22. St Peter's Church, Order of Service for the Unveiling and Dedication of the War Memorial Tablet, 5 October 1921 (Harrogate: Parish Records).
23. *LM*, 3 September 1923.
24. *Ibid.*
25. *LG*, 19 June 1860.
26. Lloyd, D.W., *Battlefield Tourism: Pilgrimage and the Commemoration of the Great War in Britain, Australia and Canada, 1919–1939* (Oxford: Berg, 1998).
27. *Ibid.*
28. Captain Francis A. Thornton, 171st Tunnelling Company RE, is buried at Mendinghem Military Cemetery in Belgium.
29. 1939 Register, RG101.
30. Name change agreed as part of the Truth and Reconciliation Commission.
31. A. Bell, Death Certificate, HMRO, 1928.
32. Board Of Trade: Outwards Passenger Lists.
33. *HA*, Friday, 10 June 1994.
34. 1939 Register, RG101.
35. *PO*, 4 November 1952.
36. *Ibid.*

37. Winter, J., *Sites of Memory, Sites of Mourning: The Great War in European Cultural History* (Cambridge: Cambridge University Press, 1995), pp.223–30.
38. Gregory, A., *The Silence of Memory: Armistice Day 1919–1946* (Oxford: Berg, 1994), pp.189–94.
39. Harvey, D. & Wallis, J. (eds), *Commemorative Spaces of the First World War Historical Geographies at the Centenary* (Abingdon: Routledge, 2017), pp.173–89.
40. Fathi, R., 'Centenary (Battlefield Tourism)', in *1914–1918: International Encyclopedia of the First World War*, ed. by U. Daniel *et al.* (Berlin: Freie Universität Berlin, 2021).
41. *HA*, 7 July 2000.
42. BBC News, 'VC Football Hero's Medals Sold at Auction', *BBC York*, 25 November 2010 (accessed 23 April 2023); available at: https://www.bbc.co.uk/news/uk-england-york-north-yorkshire-11833239.
43. Professional Footballers' Association, 'PFA Merit Award 2014: Donald Bell VC', *The PFA*, 27 April 2014 (accessed 23 April 2023); https://www.thepfa.com/news/2014/4/27/pfa-merit-award-donald-bell-vc.
44. ITV News, 'Match to Be Played in Memory of Footballer Who Won Victoria Cross', *ITV News Calendar,* 9 July 2016 (accessed 1 April 2024); https://www.itv.com/news/calendar/2016-07-09/match-to-be-played-in-memory-of-footballer-who-won-victoria-cross.
45. BBC News, 'Footballer Donald Bell: Victoria Cross Hero Remembered', *BBC News*, 10 September 2016 (accessed 23 April 2024); https://www.bbc.co.uk/news/uk-england-37318169.
46. *Ibid.*
47. Harrogate Grammar School, 'Harrogate Grammar School's Annual Celebration of Achievement', *Harrogate News*, 5 January 2016 (accessed 23 April 2025); https://www.harrogate-news.co.uk/2016/01/05/harrogate-grammar-schools-annual-celebration/.
48. Archie White letter to the editor of the *Green Howards Gazette*, Volume LXXIV.

Bibliography

Archival Sources

Commonwealth War Graves Commission
- Concentration of Grave (Exhumations and Re-Burials)
- Grave Registration Report Forms
- Headstone Reports

Green Howards Museum

Harrogate Grammar School

The National Archives
- WO 95/2167 War Diaries, 8th, 23rd and 34th Divs
- WO 95/2183 War Diaries, 24th, 68th, 69th and 70th Bdes
- WO 95/2184 War Diaries, 8th, 9th and 10th Yorks, 9th and 11th West Yorks, 10th West Ridings
- WO 95/626 War Diary of the 69th MT Coy, ASC
- WO 123/58 Army orders (War Office)
- WO 339 Soldiers Personal Documents
- WO 339 Officers Personal Documents
- WO 364 Soldiers Pension Documents
- WO 32/9394 Victoria Cross: Allowance for holders and dependants
- T 333/1 Victoria Cross including Warrants
- T 213/583 Victoria Cross: annuities payable
- CAB 45/132 to 138 Somme Authors
- BT 31/44989 and 44990 No of Company 104075; Bradford (Park Avenue) AFC Ltd.
- ED 21/20944 and 44815 Starbeck Council School
- ED 49/9066 Harrogate: St. Peter's Church of England School

North Yorkshire County Record Office
- S/HRG/8 Harrogate Municipal Secondary Day School Logbooks
- S/SBK 1/1/1 and PR/SBK 13/1 Starbeck School Principal's Logbooks and Letters
- S/HRG/9 St Peter's School Principal's Logbooks and Letters

Oxford Centre for Methodism and Church History

Westminster Training College Memorial Records

Westminster Training College Monthly War Bulletin

Westminster Training College Principal's Logbook

Western Front Association

Gunfire Archives

Stand To! Archives

TrenchMapper

Newspapers

Athletic News, Barrow Herald, Birmingham Daily Post, Bradford Daily Telegraph, The Courier, Daily Express, Daily Mirror, Daily Sketch, Durham Chronicle, Football Gazette and Telegraph, Football News, Halifax Evening Courier, Harrogate Herald, Huddersfield Daily Examiner, Hull Daily Mail, Knaresborough Post, Lancashire Evening Post, Leeds Mercury, Leicester Daily Post, Liverpool Echo, London Daily Chronicle, London Gazette, Manchester Evening News, Newcastle Chronicle, Newcastle Daily Chronicle, Newcastle Journal, Northern Echo, North Star, Pall Mall Gazette, Ripon Observer, The Scotsman, Sheffield Daily Telegraph, Sheffield Green 'Un, Sporting Life, Sports Argus, The Sportsman, Sunderland Daily Echo, Swindon Advertiser, The Times, The Weekender, Western Mail, Wharfdale & Airedale Observer, Yorkshire Evening News, Yorkshire Evening Post, Yorkshire Gazette, Yorkshire Post and Leeds Intelligencer.

Published Books and Articles

Aldrich, R., *Education for the Nation: A History of the National System of Education in England and Wales* (London: Cassell, 1996)

Alexander, J., *McCrae's Battalion: The Story of the 16*th *Royal Scots* (Edinburgh: Mainstream Publishing, 2003)

Bagwell, P., *The Railwaymen* (London: George Allen & Unwin, 1963)

Becke, A.F., *History of the 8*th *Division 1914–1918* (Aldershot: Gale & Polden, 1922)

Blunden, E., *Undertones of War* (London: R. Cobden-Sanderson, 1928)

Campbell, J., *Jutland: An Analysis of the Fighting* (London: Conway Maritime Press, 1998)

Carrington, C.E., *Soldier from the Wars Returning* (London: Hutchinson, 1965)

Crane, D., *Empires of the Dead* (London: William Collins, 2013)

Dewhirst, J., *Life at the Top* (Shipley: Bantamspast, 2016)

Douie, C.W., *The Weary Road: Recollections of a Subaltern of Infantry* (London: Heinemann, 1929)

Dunn, S., *Securing the Narrow Sea: The Dover Patrol 1914–1918* (Barnsley: Seaforth, 2017)

Easdown, M., *The History of the Metropole Hotel* (Folkestone: Local History Society, 2004)

Edmonds, J.E., *Military Operations France and Belgium, Vol. I–II (1914), Vol. I–II (1915), Vol. I (1916), Vol II (1917), Vol I–IV (1918)* (London: Macmillan & Co., 1938)

Fair, C. & R., *Marjorie's War: Four Families in the Great War 1914–1918* (Brighton: Menin House Publishers, 2012)

Falls, C., *Military Operations France and Belgium, Vol. I and III (1917)* (London: Macmillan & Co., 1938)

Fuller, J.F.C., 'The Other Side of the Hill', *The Army Quarterly* (October 1924 and January 1925)

Fussell, P., *Myth, Ritual, and Romance. The Great War and Modern Memory* (Oxford: Oxford University Press, 2013)

Gibbs, P., *The Germans on the Somme* (London: Heinemann, 1917)

Gill, D. & Dallas, G., *Mutiny at Étaples Base in 1917* (Oxford: Past & Present, 1975)

Gosse, P.H., *Memoirs of a Camp Follower* (London: William Heinemann, 1934)

Grainge, W., *Nidderdale* (Pateley Bridge: Thomas Thorpe, 1863)

Hitchen, H.S., Dawson, T.E. & Yeats, J.A., *Celebrating 150 years: The History of Wesley Chapel Harrogate 1862–2012* (Harrogate: Anchorprint, 2012)

Hartley, M. & Clapham, T., *The Avenue* (Nottingham: Temple Nostalgia Press, 1987)

Holmes, R., *Tommy: The British Soldier on the Western Front 1914–1918* (London: Harper Collins, 2004)

Holmes, R., *The Oxford Companion to Military History* (Oxford: Oxford University Press, 2001)

Inglis, S., *Soccer in the Dock* (London: Willow Books, 1985)

Jackson, A., *Football's Great War: Association Football on the English Home Front* (Barnsley: Pen & Sword Books, 2021)

Joannou, P., *To the Glory of God: Newcastle United and the Great War* (Newcastle: Novo, 2018)

Leighton, M.C., *Boy of My Heart* (London: Hodder & Stoughton, 1916)

Long, E.A., *Folkestone in the Great War* (Barnsley: Pen & Sword, 2014)

Longworth, P., *The Unending Vigil: A History of the Commonwealth War Graves Commission 1917–1984* (London: Leo Cooper, 1985)

Macdonald, L., *Somme* (London: Penguin, 1983)

Masefield, J., *The Old Front Line; or, the Beginning of the Battle of the Somme* (London: William Heinemann, 1917)

Messenger, C., *Call to Arms: The British Army 1914–18* (London: Weidenfeld & Nicolson, 2005)

Middlebrook, M., *The First Day on the Somme* (London: Allen Lane, 1971)

Miles, W., *Military Operations France and Belgium, 1916, Vol. II* (London: Macmillan & Co., 1938)

Neesam, M., *Harrogate Grammar School* (Harrogate: Manor Place Press, 2003)

O'Mara, D., *The French on the Somme: From Serre to the River Somme* (Barnsley: Pen & Sword Military, 2018)

Owen, W., *The Collected Poems of Wilfred Owen*, ed. C. Day Lewis (London: Chatto & Windus, 1963)

Plowman, M., *A Subaltern on the Somme* (London: T. Fisher Unwin, 1927)

Rawson, A., *The British Army 1914–1918* (Port Stroud: Spellmount, 2006)

Reay, S., *The Half-Shilling Curate: A Personal Account of War & Faith 1914–1918* (Solihull: Helion, 2016)

Riddoch, A. & Kemp, J., *When the Whistle Blows: The Story of the Footballers' Battalion in the Great War* (London: Haynes, 2008)

Sandilands, H.R., *The 23*rd *Division 1914–1919* (London: William Blackwood and Sons, 1925)

Shakespear, J., *The Thirty-Fourth Division* (Uckfield: Naval & Military Press, 2001)

Sheen, J., *The Green Howards in the Great War* (Barnsley: Pen & Sword Books, 2024)

Sheffield, G., *The Somme: A New History* (London: Cassell, 2003)

Simkins, P., *Kitchener's Army: The Raising of the New Armies, 1914–1916* (Manchester: Manchester University Press, 1988)

von Stosch, A., *Somme-Nord. Teil I: Die Brennpunkte der Schlacht im Juli 1916* (Berlin: Gerhard Stalling, 1927)

Sumner, I., *They Shall Not Pass: The French Army on the Western Front 1914–1918* (Barnsley: Pen & Sword Military, 2012)

Thomas, A., *A Life Apart* (London: Victor Gollancz, 1968)

Thomson, A.A., *The Exquisite Burden* (London: Jenkins Ltd, 1936)

Trapmann, A.H., *Straight Tips for 'Subs'* (London: Forster Groom Ltd, 1915)

War Office, *The Training and Employment of Grenadiers* (London: His Majesty's Stationery Office, 1915)

Winter, J., *Sites of Memory, Sites of Mourning: The Great War in European Cultural History* (Cambridge: Cambridge University Press, 1995)

Wylly, H.C., *The Green Howards in the Great War* (Richmond: [n.pub.], 1926)
Yapp, A.K., *The Romance of the Red Triangle* (London: Hodder & Stoughton, 1918)

Select Websites

Ancestry
Commonwealth War Graves Commission
English National Football Archives
Find My Past
Football and the First World War
Imperial War Museum
National Army Museum
Oxford Centre for Methodism and Church History Online Archives
The Great War Forum
The Long, Long Trail
The Western Front Association

Index